AA003601

Mathematical Models and Methods of the Analysis and Optimal Synthesis of the Developing Pipeline and Hydraulic Systems 2018

E3S Web of Conferences Volume 39 (2018)

Irkutsk, Russia
26 June – 2 July 2018

Editors:

N.N. Novitsky
M.G. Sukharev
V.A. Stennikov
V.R. Chupin
L.B. Korelstein

ISBN: 978-1-5108-6590-7

Printed from e-media with permission by:

Curran Associates, Inc.
57 Morehouse Lane
Red Hook, NY 12571

Some format issues inherent in the e-media version may also appear in this print version.

This work is licensed under a Creative Commons Attribution license:
http://creativecommons.org/licenses/by/2.0/
You are free to:
Share – copy and redistribute the material in any medium or format.
Adapt – remix, transform, and build upon the material for any purpose, even commercial.
The licensor cannot revoke these freedoms as long as you follow the license terms.
Under the following terms:
You must give appropriate credit, provide a link to the license, and indicate if changes were made. You may do so in any reasonable manner, but not in any way that suggests the licensor endorses you or your use. The copyright is retained by the corresponding authors.

Printed by Curran Associates, Inc. (2018)

For additional information, please contact EDP Sciences – Web of Conferences
at the address below.

EDP Sciences – Web of Conferences
17, Avenue du Hoggar
Parc d'Activité de Courtabœuf
BP 112
F-91944 Les Ulis Cedex A
France

Phone: +33 (0) 1 69 18 75 75
Fax: +33 (0) 1 69 28 84 91

contact-edps@webofconferences.org

Additional copies of this publication are available from:

Curran Associates, Inc.
57 Morehouse Lane
Red Hook, NY 12571 USA
Phone: 845-758-0400
Fax: 845-758-2633
Email: curran@proceedings.com
Web: www.proceedings.com

TABLE OF CONTENTS

Statement of Peer review

In submitting conference proceedings to E3S Web of Conferences, I certify to the Publisher that I adhere to the **Policy on Publishing Integrity** of the journal in order to safeguard good scientific practice in publishing.

1. All articles have been subjected to peer review administered by the proceedings editors.
2. Reviews have been conducted by expert referees, who have been requested to provide unbiased and constructive comments aimed, whenever possible, at improving the work.
3. Proceedings editors have taken all reasonable steps to ensure the quality of the materials they publish and their decision to accept or reject a paper for publication has been based only on the merits of the work and the relevance to the journal.

Title, date and place of the conference

MATHEMATICAL MODELS AND METHODS OF THE ANALYSIS AND OPTIMAL SYNTHESIS OF THE DEVELOPING PIPELINE AND HYDRAULIC SYSTEMS

26th of June – 2nd of July, 2018, Irkutsk, lake Baikal, Russia

Proceedings editor(s):

Editor-in-chief:
Novitsky N.N., Dr.Tech.Sci. Proff., Main Researcher, Head of the pipeline hydraulic systems laboratory of Melentiev Energy Systems Institute of the Siberian Branch of the Russian Academy of Science (Russia, Irkutsk).

Editors:

Sukharev M.G., Dr.Tech.Sci. Proff of Gubkin Russian State University of Oil and Gas (Russia , Moscow).

Stennikov V.A, Dr.Tech.Sci, Corresponding Member, Proff., Head of pipeline systems department, Director of Melentiev Energy Systems Institute of the Siberian Branch of the Russian Academy of Science (Russia, Irkutsk).

Chupin V.R., Dr.Tech.Sci., Proff., Head of department of Irkutsk national research technical university (Russia , Irkutsk).

Korelstein L.B., Cand.Tech.Sci. Deputy Director for scientific work of LLC "STE Pipeline" (Moscow, Russia).

Date and editor's signature

30.05.2018

Search method of the parameters network elements based on three-dimensional calculations

Dmitry V. Boykov [2,*], and *Aleksandr* A. Dekterev [1,2]

[1] Kutateladze Institute of Thermophysics SB RAS, Novosibirsk, Russia
[2] Siberian Federal University, Krasnoyarsk, Russia

Abstract. The paper proposes a method to identify a pipeline system segment, containing several piping assemblies and branches, based on three-dimensional calculations or experiments, as well as proves the applicability of proposed approach to search the local resistance coefficients of the section elements of the pipeline system, which describes the complex spatial structure, based on the known values of pressure drops and flow rates.

Introduction

Practically each industrial enterprise has one or another pipeline system in which gas or liquid flow distribution essentially influence technical, economic, and ecological characteristics of production in general. In some cases, objects under study can be represented as a hydraulic network [1]. The network may contain structural elements, whose local resistance significantly affects the nature of flow distribution. Sometimes, determining the resistance coefficients of these elements by existing empirical and analytical dependencies[2,3] is impossible.

Applying spatial simulation [4,5] to these structural elements allows defining hydraulic characteristics and using them in network design model. Earlier, the authors in [6,7] mentioned the difficulty of integrating models with a large number of independent inputs. Building a certain network analog of the element was proposed as one of the possible ways to solve this problem. Although this approach has limited applicability and accuracy as compared to the option with direct integration of the three-dimensional element into the network model [8], it allows quickly constructing the hydraulic network model and obtaining the results of calculations.

As an example of the use of proposed approach, the authors consider the problem on simulation of a multi-tiered scrubber (Fig.1). Scrubber is a complex object with five independent inlets. Using a full spatial model or hybrid model with directly embedded three-dimensional element into a network model [6,7] results in excessive computational costs. The authors used a simplified network model of this object (Fig. 2). To determine the parameters of the network elements, several three-dimensional calculations were carried out.

* Corresponding author: dimkadim@yandex.ru

In consequence of these calculations, knowing flow rates at the inlet, we determined pressure drops between the scrubber inlets and outlets (Table. 1).

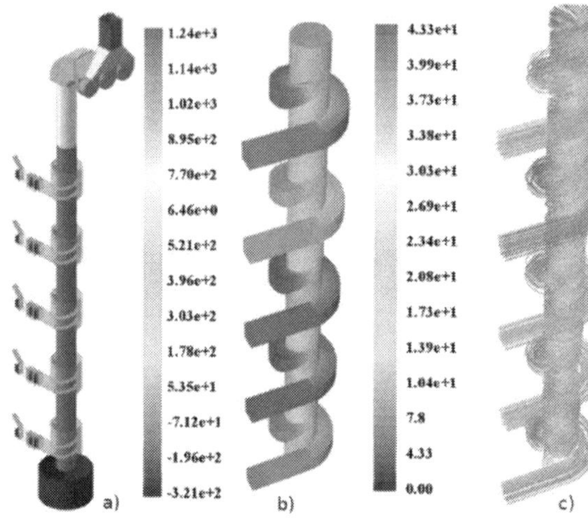

Fig. 1. Scrubber: a) geometry; b) static pressure (Pa); c) streamlines.

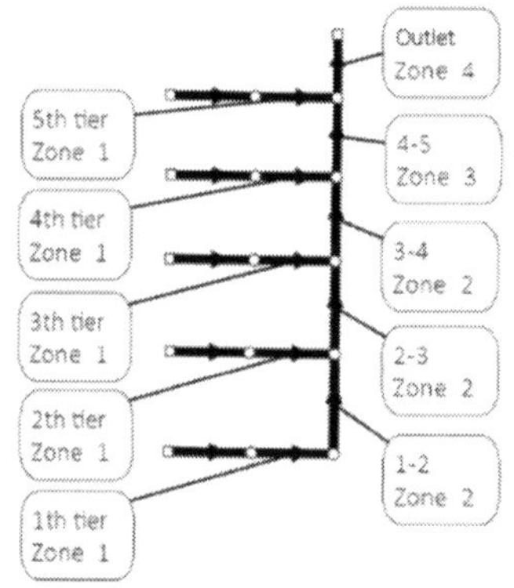

Fig. 2. Network model of scrubber.

The network analogue of the spatial scrubber model (Fig. 2) was divided into several zones:

1. the zone of all inlets, whose geometric similarity suggests the existence of a single local resistance coefficient at all sites in the zone;
2. the zone of a lifting pipe from the 1st to 4th tier, where small velocities allowed suggesting that the influence of the local resistance of these tiers on the flow distribution in the scrubber will be insignificant and, accordingly, local resistance was set to 0 for all sections of the zone;
3. the zone of a lifting pipe from the 4th to 5th tier;
4. the zone of access to the fan.

Accordingly, the problem was reduced to finding three local resistance coefficients for zones 1, 3 and 4 satisfying the condition of minimum difference between the pressure drop in the scrubber for network computations and 3D calculations. Initially, the coefficients were found by the exhaustive search method [6,7]. Unfortunately, such a search method becomes too expensive at a large number of unknown parameters. In this regard, it turned out to be advisable developing a search algorithm to find parameters of the network section based on known data from the results of 3D calculation or experiment.

Table 1. Calculation data of scrubber 3D model

	1st 3D calculation		2nd 3D calculation		3rd 3D calculation	
	Pressure, Pa	Flow rate, kg/s	Pressure, Pa	Flow rate, kg/s	Pressure, Pa	Flow rate, kg/s
1st tier	799	6.60	2712	13.20	555	6.60
2nd tier	952	7.80	2019	7.80	210	0.10
3rd tier	950	8.12	3263	16.30	196	0.10
4th tier	656	5.14	1529	5.14	383	5.14
5th tier	934	9.86	1566	9.86	836	9.86
Outlet	0	37.52	0	52.27	0	21.80

Flue ducting calculation model

As a mathematical model to describe the fluid flow in the network, a model of steady-state flow of incompressible fluid was adopted. For representation of pipeline system we use the directed graph[9], whose constraint matrix is presented in the following form:

$$\nabla_{il} = \begin{cases} 1, \text{ if } l \in O_i, \\ -1, \text{ if } l \in I_i, \\ 0 \quad otherwise \end{cases} \tag{1}$$

where $l \in O_i$– is the plurality of pipes coming out from the *i*-th piping assembly; $l \in I_i$– is the plurality of pipes going into the *i*-th piping assembly. The problem of flow distribution in the network can be reduced to a combination of the mass conservation law in the piping assembly (2), and the flow resistance law in the pipe (3):

$$\sum_{l \in U_j} \nabla_{jl} x_l = 0 \tag{2}$$

$$\sum_{i \in N} \nabla_{il} p_i = h_l\left(x_l\right) \tag{3}$$

where N – is the set of all network piping assemblies; U – is the set of all network branches; x_l – is the flow rate in the branch (kg/s); $h_l(x_l)$ – is the pressure loss (Pa); p – is the pressure (Pa).

Finding a solution to this nonlinear system of equations is carried out by the nodal pressure method. That is, the following flow rate and pressure is sought through the correction:

$$x_l^{K+1} = x_l^K + \delta x_l \tag{4}$$

$$p_i^{K+1} = p_i^K + \delta p_i \tag{5}$$

Substituting equations (4) and (5) in (2) and (3) it is easy obtaining a system of linear algebraic equations (SLAE) for the pressure correction:

$$\begin{aligned} &\sum_{i \in N} \Delta_{ij}\left(a\right)\delta p_i = -\sum_{l \in U_j} \nabla_{il} x_l, \, j \in N_{inner} \\ &\delta p_j = 0, \; j \in N_{bdr} \end{aligned} \tag{6}$$

$$\Delta_{ii}(a) = \sum_{l \in U_i} a_l, \; i \in N_{inner}$$

$$\Delta_{ij}(a) = -\sum_{l \in U_i} a_l, \; i \in N_{inner}, j \in N, i \neq j$$

$$a_l = \begin{cases} \left(\dfrac{\partial h_l}{\partial x_l}\right)^{-1}, l \in U_{inner} \\ 0, l \in U_{bdr} \end{cases}$$

here N_{bdr}, N_{inner} – are sets of boundary nodes and computational nodes, respectively; U_{bdr}, U_{inner} – are sets of boundary branches and computational branches, respectively. Pressure is fixed in the boundary nodes, while flow rate is fixed in the boundary branches.

Search for local resistance coefficients

For the identified branches, we determine the correction of the analyzed value in the form of [10]:

$$C^{K+1} = C^K + dC \tag{7}$$

where C can be, for example, diameter or local resistance.

Substituting (5) and (7) in (2) and (3) we obtain SLAE, which associates the correction of some magnitude in the branches with the resulting pressure change in the nodes.

$$\begin{gathered} \sum_{i \in N} \Delta_{ij}(a) dp_i = \sum_{l \in U} \nabla_{il} \left(\frac{\partial h_l}{\partial x_l}\right)^{-1} \left(\frac{\partial h_l}{\partial C}\right) dC, \; j \in N_{inner} \\ dp_j = 0, \; j \in N_{bdr} \end{gathered} \tag{8}$$

Where dp_i- is the pressure change; dC - is the correction. The left part of equation (6) at the final iteration coincides with the left part of the SLAE of the correction (8).

It is easy to show that the resulting pressure change for the sum of the different impacts is equal to the sum of the pressure changes of these impacts. Using the computation results of the SLAE (8) this allows determining the SLAE for correction of the variable:

$$\frac{dp_I^m}{dC_m} \cdot \delta C_m = P_I - p_I, \; m \in U_{ident}, I \in M_{control} \tag{9}$$

here δC_m – is the correction of the identified value, $m \in U_{ident}$ – is the set of identifiable objects, P_I^m – is the required pressure in the control points, p_I^m – is the pressure in the specified nodes (control points) of the network model for which the model is adapted, $I \in M_{control}$ - is the set of control points.

The coefficient dp^m_I/dC_m is defined as the ratio of pressure change in the control node dp_I to the corresponding change of the identified value $dC_m$and is found through the solution of the equations system (8) for all options of control actions. The object being identified can be understood as both a single branch and a group of similar branches (by example 1, Fig. 2).

In this example, the control points are the inlets to the scrubber (Table 1). The required pressure P_I is taken from the corresponding 3D calculation (three 3D calculations, 5 nodes in each, 15 calculation points in total). The resulting system of equations is over-defined [11]. A fixed flow rate is given on boundary branches.

The coefficient of local resistance in the formula for pressure loss is the identifiable value: $h_l(x_l, \xi_m) = \xi_m \frac{|x_l| \cdot x_l}{2 \cdot \rho_l \cdot s_l^2}$ where ξ_m – is the local resistance coefficient; s_l – is the cross-sectional area of the pipeline (m^2), and ρ_l – is the flow density (kg/m^3). The right part of SLAE (8) can be represented as:

$$\left(\frac{\partial h_l}{\partial x_l}\right)^{-1}\left(\frac{\partial h_l}{\partial \xi_m}\right) d\xi_m = \frac{x_l}{2\xi_m} d\xi_m . \quad (10)$$

The above allows proposing the following algorithm for searching the local resistance of identifiable branches: The network model (2, 3) is calculated for each three-dimensional calculation, in which the flow rate is fixed at the boundary branches. To calculate matrix columns $dp^m{}_l/dC_m$, SLAE (8) is solved for each three-dimensional calculation and for each identifiable object. New local resistance coefficients are calculated after solving resulting SLAE (9), after which the cycle repeats over again.

The result of solving the test task

Table 2. The local resistance coefficient

	Search method	Option 1	Option 2	Option 3
1st tier	2.8	2.4	2.5	2.7
2nd tier	2.8	2.0	2.5	2.7
3rd tier	2.8	2.3	2.5	2.7
4th tier	2.8	2.2	2.5	2.7
5th tier	2.8	2.6	2.5	2.7
1st and 2nd tiers	0	0	1.8	0
2nd and 3rd tiers	0	0	0	0
3rd and 4th tiers	0	1.5	1.2	0
4th and 5th tiers	2.8	2.1	1.8	2.5
Outlet	2	2.2	2.3	2.1

The proposed algorithm was applied to calculate the multi-tier scrubber. Three grouping options of local resistance coefficients were considered: 1– all branches are considered to be independent; 2 – inlets at tiers are united into a single block (zone 1 in Fig. 2); and 3 – the option fully corresponds to the problem statement as in the exhaustive search method. Table 2 shows the local resistance coefficients for all three options. The exhaustive search method and the method proposed in option 3 gave similar results.

Table 3 shows the comparison of the models with respect to the pressure drop. In all options, pressure drops are close to the data obtained through 3D modeling, while the error does not exceed the accuracy of calculations accepted for engineering practice.

Conclusion

We proposed a method for determining the parameters of the network branches based on to the three-dimensional calculation. The local resistance coefficients of several branches of the network model based on the spatial simulation are determined on the example of calculations of a multi-tier scrubber.

Table 3. Comparison of network models

No. of 3D caculation			1st calculation		
	3D calculation	**Exhaustive search method**	**Option 1**	**Option 2**	**Option 3**
1st tier	799	990	1003	995	966
2nd tier	952	1118	1053	1095	1086
3rd tier	950	1154	1113	1127	1121
4th tier	656	854	778	788	837
5th tier	934	1087	1003	1059	1076
No. of 3D calculation			**2nd calculation**		
	3D calculation	**Exhaustive search method**	**Option 1**	**Option 2**	**Option 3**
1st tier	2712	2714	2680	2678	2619
2nd tier	2019	1817	1915	1901	1819
3rd tier	3263	3385	3215	3221	3255
4th tier	1529	1595	1454	1446	1556
5th tier	1567	1431	1459	1464	1447
No. of 3D calculation			**3rd calculation**		
	3D calculation	**Exhaustive search method**	**Option 1**	**Option 2**	**Option 3**
1st tier	555	502	460	481	489
2nd tier	210	179	189	188	182
3rd tier	196	178	188	188	181
4th tier	383	375	332	351	367
5th tier	836	844	800	773	816

References

1. A.P. Merenkov, V.Ya. Khasilev *Theory of hydraulic circuits*. (1985)
2. I. E. Idel'chik, , *Handbook of Hydraulic Resistance*. (1992)
3. D. S. Miller, *Internal flow systems*. (1990)
4. J.H. Ferziger, M. Peric, *Computational Methods for Fluid Dynamics*. (2002)
5. A.A. Dekterev, K.Yu. Litvintsev, A.A. Gavrilov, E.B. Kharlamov, S.A. Filimonov Journal of Siberian Federal University. Engineering & Technologies, **10(4)** (2017)
6. D. V. Boykov, A. A. Dekterev, S. A. Filimonov. Pipeline energy systems. Mathematical and computer technologies of intellectualization., pp 138-151 (2017)
7. D. V. Boykov, S. A. Filimonov, A. A. Dekterev,. Mathematical modeling of pipeline energy systems. Proceedings of the XIV all-Russia Scientific Seminar with International Participation "Mathematical models and methods of analysis and optimal synthesis of developing pipeline and hydraulic systems" pp. 28-43 (2014)
8. S.A. Filimonov, E.I. Mikhienkova, A.A. Dekterev, D.V. Boykov, Journal of Physics: Conference Series, **Vol 899** (2017)
9. *R.* Wilson. *Introduction to graph theory* (1996)
10. S.P. Epifanov, N.N. Novitsky. Pipeline energy systems. Methods of mathematical modeling and optimization. pp 27-47 (2007)
11. Benoît Jacob, Gaël Guennebaud. **Eigen 3 documentation**

Probabilistic modeling of hydraulic conditions of tree pipeline networks under conventional boundary conditions*

Olga V. Vanteeva [1,†]

[1]Melentiev Energy Systems Institute of Siberian Branch of the Russian Academy of Sciences (ESI SB RAS), Pipeline Energy Systems Department, 130, Lermontov str., Irkutsk, Russia, 664033

Abstract. The paper is concerned with the probabilistic modeling of pipeline network hydraulic conditions that occur due to random impacts of the external environment (load of consumers, pressures at sources). The research is focused on a network with a tree configuration under conventional boundary conditions (for pressure at a tree root and flow rates at the other nodes) that are specified probabilistically. We propose a topological method for a probabilistic analysis of such network operation. This method makes it possible to determine mathematical means, variances and covariances of state variables (pressures at nodes, flow rates and pressure losses in branches) by final formulas when traversing the entire tree without laborious matrix operations or calculations of systems of equations. A numerical case study is used to exemplify the efficiency of the proposed approach, adequacy of the obtained results and high speed compared to other, more general methods similar to Monte-Carlo methods.

1 Introduction

Flow distribution problems are primary in an analysis and substantiation of operating conditions of pipeline systems of various types and purposes for their design, operation and dispatch control. Traditionally, these problems are solved with deterministic mathematical models and methods, which however do not allow the assessment of the extent to which the pipeline system operation affected by a set of random factors is uncertain. This explains the relevance of the probabilistic flow distribution problem statements, which imply obtaining calculation results in the form allowing probabilistic interpretation based on the information on the so called boundary conditions in a probabilistic form. Solving such problems with traditional methods, for example with Monte Carlo method [1], is associated with excessive computational efforts even for the pipeline systems of a medium size.

2 Analysis of literature and problem statement

To date, Russia and other countries have carried out many studies dealing with the tasks

*The research was carried out within the project III.17.4.3 of the Fundamental research program of SB RAS (AAAA-A17-117030310437-4)
† Corresponding author: vanteeva@isem.irk.ru

and methods for calculation of steady-state flow distribution in pipeline systems of different types and purposes.

The experience of flow distribution modeling in the papers by E. Todini [2–4], O. Giustolisi [5,6], is characterized mainly by their focus on deterministic, steady-state and quasidynamic flow distribution. In Russia, the flow distribution problems similar in statement were solved already in the 1930s by Andriyashev M.M. [7], and Lobachev V.G. [8], and generalized in the studies by A.P.Merenkov and V.Y.Khasilev [9].

The issues of stochastic flow distribution and optimization of parameters under uncertainty in water supply systems (in their design) are considered in the studies by S.N.Karambirov [10, 11]. In these studies, the statistical models (time series, autocorrelation functions) are used to take into account the stochastic nature of water consumption. The authors of [12, 13] put an emphasis on probabilistic problems considered in the analysis of pipeline system operating conditions, given pumps and reservoirs. These problems are mainly solved by Monte Carlo method which is also widely used in the analysis of pipeline system reliability.

The authors of [14–18] propose an approach to the development of analytic probabilistic models of steady-state hydraulic conditions. This approach, for example, can be applied to the water and heat supply systems under stochastic water consumption. The general and stage-by-stage methods presented in [14-17] are characterized by their universality, satisfactory accuracy and low computational efforts compared to the classical Monte Carlo method. They can be used to determine all probabilistic characteristics of steady-state conditions in pipeline systems of random configuration and any set of boundary conditions specified in a probabilistic form. The methods are aimed at obtaining mathematical means for state variables and calculating a covariance matrix of these variables based on a specified covariance matrix of boundary conditions and a derivative matrix of flow distribution model at the point of mathematical mean of boundary conditions, and involve matrix operations including inverse of a derivative matrix [14-18].

3 Object, goal and objectives of studies

An object of the research discussed in the paper is a tree pipeline network. The goal is to develop an effective method and an algorithm for calculation of statistic characteristics (mathematical means, variances, and covariances) of hydraulic state variables (pressures, flow rates) that occurred as a result of random impact of the external environment and that are specified probabilistically for pressure at one node (a tree root) and flow rates at the other nodes.

To achieve this goal we set the following objectives for the research: 1) to obtain final formulas for calculation of statistic characteristics of state variables; 2) to develop a topological method (and an algorithm for calculation), that identify the sequence in which the obtained formulas should be applied; 3) to numerically check the operability of the topological method and accuracy of the obtained solution in comparison with the other methods (Monte Carlo and general methods).

4 The main principles of the topological method

We will introduce the notion of a tier of nodes with index *k*, which represents a set of nodes J_k, such that $|R_j|=k$, where R_j – a set of branches belonging to the path from a tree root to node *j*. Thus, the *k*-th tier contains all nodes of the scheme that are connected with root *k* by branches. Let us put the direction of branches of the tree scheme in correspondence with the direction of flows. Assign the first numbers (indices) to the nodes of the last tier, the next numbers - to the tier before last, and so on. Only one flow goes into each node, therefore the branches can always be numbered so that the branch index will coincide with the index of a final node, i.e. $i = \text{f}_i = j$. Let us introduce a set of branches I_j, incident to node j, that can now be represented by $I_j = j \cup I_j^-$, where I_j^- is a set of branches going from node j.

By virtue of the mentioned principle of node and branch numbering, we obtain:

1) $\text{o}_i > \text{f}_i$, $i = \overline{1,n}$, where o_i, f_i are the indices of initial and final nodes of the i-th branch;

2) node with an external source (tree root) will have index m;

3) $J_0 = \{m\}$, $J_{k+1} = \bigcup\limits_{j \in J_k} I_j^-$, $k = \overline{1, K-1}$;

4) the number of tiers in the scheme will be determined as $K = |R_1| + 1$;

5) $j > l$ for all $j \in J_k$ and $l \in J_{k+1}$, $k = \overline{0, K-1}$, therefore, the enumeration of nodes $j = 1,...,m$ means a successive traversing through the tiers in the direction of $k = K-1,...,0$ and on the contrary, enumeration of nodes $j = m,...,1$ implies traversing in the direction of $k = 0,...,K-1$. Figure 1 demonstrates an example of introduced numbering.

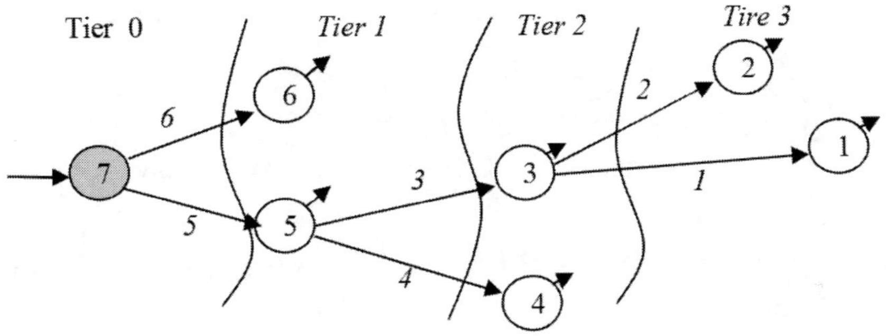

Fig.1. An example of a tree scheme numbered for calcualtion by the topological method.

The topological method implies successively traversing the nodes of the calculation scheme by tiers, starting with the nodes for which the boundary conditions are specified with relatively simple algebraic relationships in each step to determine the unknown values of mathematical means, variances and covariances of state variables.

From now on we assume that the set of boundary conditions (vector G) is conventional $G = (Q_1, Q_2, ..., Q_{m-1}, P_m)^{\text{T}}$, where Q_j, P_j are flow rate and pressure at the *j-th* node. Boundary conditions are specified by their mathematical means $\overline{G}$ and covariance matrix C_G. Without loss of generality, for simplicity we will also consider all boundary conditions to be mutually uncorrelated, when $\text{cov}(G_j, G_l) = 0$ for $j \neq l$ and $\text{cov}(G_j, G_l) = \sigma_j^2$ for $j = l$, $j, l = \overline{1,m}$, i.e. $C_G = \text{diag}(\sigma_{Q1}^2, \sigma_{Q2}^2, ..., \sigma_{Qm-1}^2, \sigma_{Pm}^2)$ is a diagonal matrix of order $\dim(G) = m$.

5 Calculation of mathematical means of state variables

Let us explain the principle of traversing the nodes of the scheme on the example of a calculation of mathematical means of state variables.

Linearly independent equations of the first Kirchhoff law [9] will be presented in an index form: $x_j - \sum_{i \in I_j^-} x_i = Q_j$, $j = \overline{1, m-1}$, where x_i is flow rate in the i-th branch. The mathematical means of flow rates in the branches are calculated successively by the expression

$$\bar{x}_i = \sum_{t \in I_i^-} \bar{x}_t + \bar{Q}_i, \ i = \overline{1,n}, \tag{1}$$

which takes into account the fact that $i > t \in I_i^-$. Therefore, by the time of calculating $\bar{x}_i$, all $\bar{x}_t$, $t \in I_i^-$ are already known.

The pressure drop in branches $y_i = f_i(x_i)$, $i = \overline{1,n}$, where $f_i(x_i)$ is function for hydraulic characteristics of the branches, therefore, (1) can be used to calculate

$$\bar{y}_i \approx f_i(\bar{x}_i), \ i = \overline{1,n}. \tag{2}$$

The remaining linearly dependent equation of the first Kirchhoff law makes it possible to determine

$$\bar{Q}_m = \sum_{i \in I_m^-} \bar{x}_i. \tag{3}$$

The mathematical means of nodal pressures are calculated using an index analog of equations of the second Kirchhoff law in a nodal form $P_{\text{o}i} - P_{\text{f}i} = y_i$, hence

$$\bar{P}_j = \bar{P}_{\text{o}j} - \bar{y}_j, \ j = \overline{n,1}. \tag{4}$$

The calculation starts for the nodes of tier $k = 1$, when $\text{н}_j = m$, and $\bar{P}_m$ is known. In calculation of $\bar{P}_j$ for $k > 1$, the values of $\bar{P}_{\text{o}j}$ are also known, because $\text{o}_j > j$.

6 Index-recurrent relationships for variances and covariances of state variables

We will present the final relationships for the variances of state variables [18]:

1) for the variance of flow rates in branches

$$\sigma_{x,i}^2 = \sigma_{Q,i}^2 + \sum_{t \in I_i^-} \sigma_{x,t}^2, \ i = \overline{1,n}; \tag{5}$$

2) for the variance of pressure drops in branches

$$\sigma_{y,i}^2 = \left(f_{x,i}'\right)^2 \sigma_{x,i}^2, \ i = \overline{1,n}, \tag{6}$$

where $f_{x,i}' = \partial f_i / \partial x_i$, $i = \overline{1,n}$;

3) for the variance of flow rates in branches

$$\sigma_{P,j}^2 = \sigma_{P,\text{H}j}^2 + \sigma_{y,j}^2 + 2f_{x,j}'\sigma_{x,j}^2 f_{\Sigma,\text{H}j}, \ j = \overline{m-1,1}; \tag{7}$$

The value of $f_{\Sigma,i}$ is calculated recurrently –

$$f_{\Sigma,m} = 0, \ f_{\Sigma,j} = f_{\Sigma,\text{H}j} + f_{x,j}', \ j = \overline{m-1,1}; \tag{8}$$

4) for the variance of the nodal flow rate at the source

$$\sigma_{Q,m}^2 = \sum_{t \in I_m^-} \sigma_{x,t}^2. \tag{9}$$

7 Topological algorithm

Thus, the proposed "topological" method for probabilistic modeling of hydraulic conditions for a tree network can be reduced to the following computational scheme.

1. Forward pass. For each $i=\overline{1,n}$, calculate: $\bar{x}_i$, $\sigma^2_{x,i}$ with respect to (1), (5), and $\bar{y}_i$, $\sigma^2_{y,i}$ with respect to (2), (6). Determine $\bar{Q}_m$, $\sigma^2_{Q,m}$ with respect to (3), (9).

2. Backward pass. By assuming $f_{\Sigma,m}=0$, for each $j=\overline{m-1,1}$, determine $\bar{P}_j$ and $\sigma^2_{P,j}$ with respect to (4), (7), and $f_{\Sigma,j}$ - with respect to (8).

8 Numerical case study

Let us compare the probabilistic flow distribution characteristics for the scheme in Figure 1, that were calculated by different methods: Mote Carlo method, general method and topological method. The boundary conditions represent a set of mutually uncorrelated magnitudes $G=\{Q_1,...,Q_6,P_7\}$ with numerical values of mathematical means $\bar{G}=\{50;\ 5;\ 5;\ 3;\ 8{,}06;\ 5;\ 7\}$ and mean square deviations $\{0{,}1;\ 0{,}86;\ 0{,}87;\ 0{,}3;\ 0{,}887;\ 0{,}5;\ 1{,}264\}$. The functions for hydraulic characteristics of branches have the form $f_i(x_i)=s_i\,|x_i|\,x_i$, $i=\overline{1,n}$, where s_i – hydraulic resistance of the i-th branch. These resistances have the values {0,015; 0,0908; 0,0108; 0,015; 0,00646; 0,00557}. Consequently, $\partial f_i/\partial x_i=2s_i\,|x_i|$, $i=\overline{1,n}$.

Tables 1 and 2 contain calculated probabilistic characteristics for flow rates in the branches and nodal pressures by different methods. The characteristics of the nodal flow rate of the sources found by different methods are: $\bar{Q}_7$ =33,07, $\sigma^2_{Q,7}$= 2,046 (MCM); $\bar{Q}_7$ =33,06, $\sigma^2_{Q,7}$= 2,054 (GM and SSM).

Table 1. Results of probabilistic modeling of flow rates in branches

Branch, i	Monte Carlo method (N=5000)		General method		Topological method	
	$\bar{x}_i$	$\sigma_{x,i}$	$\bar{x}_i$	$\sigma_{x,i}$	$\bar{x}_i$	$\sigma_{x,i}$
1	4,99	0,86	5	0,86	5	0,86
2	4,99	0,88	5	0,87	5	0,87
3	12,99	1,27	13	1,26	13	1,26
4	8,06	0,88	8,06	0,88	8,06	0,88
5	26,04	1,61	26,06	1,62	26,06	1,62
6	7,01	1,26	7	1,26	7	1,26

Table 2. Results of probabilistic modeling of nodal pressures

Node, j	Monte Carlo method (N=5000)		General method		Topological method	
	$\bar{P}_j$	$\sigma_{P,j}$	$\bar{P}_j$	$\sigma_{P,j}$	$\bar{P}_j$	$\sigma_{P,j}$
1	43,40	0,94	43,41	0,94	43,41	0,94
2	41,51	1,49	41,51	1,48	41,51	1,48
3	43,77	0,85	43,78	0,85	43,78	0,85
4	44,62	0,69	44,63	0,69	44,63	0,69
5	45,62	0,55	45,61	0,55	45,61	0,55
6	49,72	0,14	49,72	0,14	49,72	0,14
7	50,00	0,1	50,00	0,1	50,00	0,1

As is seen, the results obtained by different methods virtually coincide. The calculation time with the topological method, however, is many times less than with general and Monte Carlo methods: Monte Carlo method (5000 realizations) – 206 s; General method – 1 s.; Topological method – 0.25 s.

9 Conclusions

The probabilistic modeling of flow distribution in the pipeline system of a tree configuration has been studied. The final expressions for the mathematical means and variances of state variables are calculated recurrently. A new topological method and an algorithm are developed to calculate all statistic parameters of hydraulic conditions while traversing the tree. The proposed method enables calculations without laborious matrix operations or solving the systems of equations. The numerical cases are used to exemplify the adequacy of the results obtained with the proposed method and its greater computational efficiency compared to the other alternative methods.

References

1. I.M. Sobol. *Numerical Monte Carlo methods* (Science, Moscow, 1973) [in Russian].
2. Todini, E. Bollettinodegli Ingegneridella. Toscana **11** (1979)
3. Todini E., Pilati S. *A gradient algorithm for the analysis of pipe networks.* (L., John Wiley & Sons, 1988)
4. Todini, E., Computing and Control for the water Industry (1999)
5. O. Giustolisi, Z. Kapelan, D. Savic, J. of water resources planning and management. **134,** 6 (2008)
6. O. Giustolisi. Proc. of WEX Workshop, Marbella, Spain. 2008. (CD-ROM)
7. M.M. Andriyashev. *A technique for calculation of water pipeline networks* (M., Soviet Legislature, 1932)
8. V.G. Lobachev. Santekhnika. **2**, (1934)
9. A.P. Merenkov and V.Ya. Khasilev, *The Theory of Hydraulic Circuits* (Science, Moscow, 1985)
10. S.N. Karambirov. *Mathematical modeling of systems for supply and distribution of water under a great number of operating conditions and uncertainty* (M., MGUP, 2004)
11. Karambirov S.N. Abstract of PhD Thesis (M., 2005)
12. H.A. Kretzmann, J.E. van Zyl and J. Haarhoff, Proceedings of the 2004 Water Institute of Southern Africa (WISA) (Biennial Conference 2 – 6 May 2004)
13. P. Cutore, A. Campisano, Z. Kapelan, C. Modica, D. Savic. Urban Water Journal, **5,** 2 (2008)
14. N.N. Novitsky, O.V. Vanteeva. Saint Petersburg Polytechnic University Journal of Engineering Science and Technology, **1** (2008)
15. N.N. Novitsky, O.V. Vaneeva, Proceedings of the Russian Academy of Sciences. Power Engineering, **2** (2011)
16. O.V. Vanteeva. Abstract of PhD Thesis (Irkutsk ESI SB RAS, 2011)
17. N.N. Novitsky, O.V. Vanteyeva. Chaotic Modeling and Simulation (CMSIM), **1** (2014)
18. N.N. Novitsky, O.V. Vanteeva, *Proceedings of the Russian Academy of Sciences. Power Engineering*, **6** (2017)

Generalization of multi-level modeling methods for development and analysis of operating conditions of large heat supply systems*

Zoya. I.Shalaginova[1], and *Vyacheslav* V.Tokarev[1]

[1]Melentiev Energy Systems Institute of Siberian Branch of the Russian Academy of Sciences (ESI SB RAS), Pipeline Energy Systems Department, 130, Lermontov str., Irkutsk, Russia, 664033

Abstract. The paper presents a methodology of multi-level modeling of thermal-hydraulic conditions of large heat supply systems on the basis of the scientific area of research – the theory of hydraulic circuits – that has been developed at MESI SB RAS. The essentials of the applied methods and also the mechanism of their integration with the up-to-date information technologies are described. In combination they make it possible to calculate operating conditions of heat supply systems (HSS) of arbitrary size and structure. The applied approach is based on the multi-level arrangement of data and single-and multi-level calculations. The latter, in turn, are based on application of the methods of equvalenting and decomposition of both the calculation schemes and problems. The methods for multi-level adjustment thermal-hydraulic calculations take into consideration all requirements for feasibility of conditions and are realized as the information-computing system (ICS) "ANGARA-TS". The problems to be solved using the ICS and the technology for development of operating conditions are presented. The ICS is applied to sets of real HSS of big towns to develop conditions and adjustment measures. The composition of the solved tasks implemented in the ICS, the technology of development of operating conditions and an example of a multilevel modeling of the thermal-hydraulic conditions of the HSS in Petropavlovsk-Kamchatsky are presented.

1. Introduction

The Russian heat supply systems are engineering and technical facilities that are unique in their scales and complexity and an essential part of the energy sector, industry, municipal sphere and other sectors of the national economy. The systems are complex spatial large objects which combine a set of heterogeneous components, develop in time and are distributed over the substantial territory. Availability of diverse heat sources (cogeneration plants, district and peak boiler plants), pumping stations, heat points – central and individual, other control units, numerous diverse consumers with different schemes of load connection, extended and looped heat networks with ever-changing operation conditions of HSS makes the problems of development of conditions and their control complex.

The main causes of change in operation conditions of HSS are: inconstancy of their structure in the processes of development, reconstruction and connection (disconnection) of consumers, scheduled maintenances and emergency situations; variability of parameters due to natural aging and wear of equipment, operation of automatic devices; external random effects such as ambient temperature variations, interruptions in operation of adjacent systems (electric power, fuel supply, water pipeline), variability of uncontrolled load of hot water supply (HWS), etc.

*The study was carried out within the framework of the III.17.4.3 project of the Basic Research Program of the SB RAS (AAAA-A17-117030310437-4).

The governing factor in maintaining the quality of heat supply to consumers with reduction of operating costs is a smart planning of operation conditions of HSS, which is possible only on the basis of preliminary calculations to be made. Traditionally the problems of planning normal and post-emergency conditions of HSS are solved by the multi-variant calculations of flow distribution in advance, before the heating season. Moreover, change in operation conditions within the heating period can lead to the fact, when the operating parameters exceed technologically acceptable limits and as a consequence, to equipment damage or its faster wear, violation of heat supply to consumers, etc. This necessitates correction of conditions with change in operational environment.

In addition to large dimensionality and compound structure of large HSS, complexity of their modeling is conditioned by heterogeneity of elements, which require a different degree of detail for solving the operation problems. The multi-level modeling [1–6] developed at MESI SB RAS is a way to overcome contradiction between high dimensionality and requirements of integrity and visualization of the studied object. The approach is based on integration of models and methods for calculation of conditions with information technologies. In combination, they make it possible to link up scattered and inconsistent initial data with the models, solving thereby laborious problems of the quantitative validation of solutions obtained.

Irrespective of the methods of modeling, large HSS have a hierarchical structure to provide multi-stage control: at heat sources; at intermediate stages – central or individual heat points; and directly in heating systems, ventilation systems and hot water supply systems. As a rule, heat networks in such systems have different levels – main and distributionи networks with the central heat points serving as an interface. This fact creates natural opportunities for using multi-level modeling to calculate conditions of large HSS. Here the conditions of main and distribution networks can differ in the temperature charts of heat supply and the method of control – quantitative and qualitative.

2. Basic principles of multi-level modeling

The multi-level modeling technology enables to: 1) decrease dimension of the problems solved; 2) visualize calculation results; 3) coordinate solution results of the problems requiring different detail of the models; 4) overcome disconnection of information and mathematical models for solving the HSS control problems at different departmental, territorial, organizational and time levels; 5) apply technologies of parallel calculations, which in turn enables to calculate high-dimensional HSS for a shorter time.

The multi-level modeling can be realized by applying multi-level arrangement of data and arrangement of single- and multi-level calculations. The multi-level calculations are based on application of the methods of equivalenting and decomposition of both calculation schemes and problems [1, 4, 5]. This makes the applied calculation methods independent of the level of scheme aggregation and detail.

The decomposition principles of calculation schemes are described in [1, 4]. According to these principles main networks include networks from heat sources to central heat points (if they are absent – to the points of connection to the residential districts). Besides, the main networks comprise pumping stations and sections of the distribution networks, if they form loops with the main networks. The dead ends of central heat points to consumers also refer to distribution networks. The distribution network of one district can consist of several independent fragments, whose number is equal to the number of the points of connection to the residential district. The central heat point with installed heat exchangers in the closed heat supply systems is the common element that transmits parameters to coordinate conditions of different levels. In the open systems the common element is a generalized consumer at the main network level with the total load of the residential district, which is a generalized source at the distribution network level. The calculation schemes must be decomposed in terms of possible calculation of the whole heat supply system with mutual coordination of networks of different levels. Correct generation of the calculation scheme determines effectiveness of modeling. The main criteria of scheme decompositions are the degree of the impact of HSS elements on operation conditions of the whole system; possibility of calculation and analysis of the condition after control actions; possibility of operational control of change in the scheme configuration. Accordingly, the following structure of the calculation scheme decomposition is developed (Fig. 1).

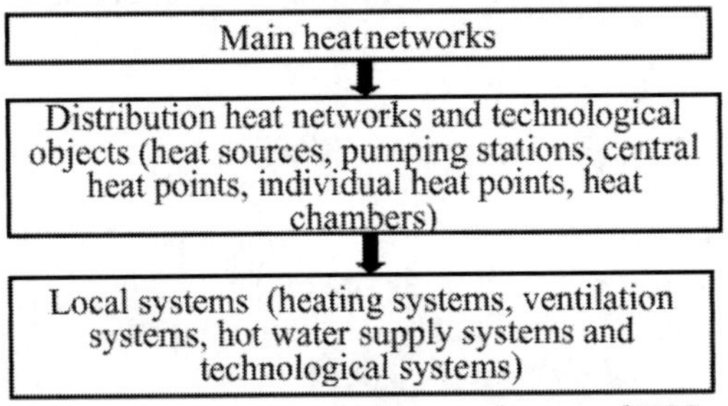

Fig. 1. Hierarchy of calculation schemes of HSS.

Such principles of scheme decomposition of large HSS with many thousands of elements make it possible to consider different departmental networks of different levels, visualize the studied object and perform modeling for various purposes. The mechanisms for supporting the hierarchy of calculation schemes are presented in [3].

The principles of problem decomposition are described in [1, 4]. The assumptions made allow the calculation of hydraulic condition irrespective of thermal condition and decomposition of the problem on thermal-hydraulic calculation of HSS into two subproblems: the hydraulic-thermal condition of the network and the hydraulic-thermal condition of the heat point. Each subproblem is also reduced to solution of the problem of hydraulic condition calculation and the thermal condition calculation of its base. The issues of modeling the hydraulic and thermal conditions of heat networks were analyzed in the works [7-13]. The mathematical model of thermal-hydraulic conditions of heat points is described in [14-15].

The calculation scheme of any HSS object can be represented by one element in the scheme of a higher level. The equivalenting procedure is performed for this element in advance, i.e. the parameters of aggregated characteristics, which are assigned to this element and used to calculate the scheme of an upper level, are calculated. Then the procedure of disaggregation, which consists in specification of the results of calculating an upper level element, is performed for all elements of its own scheme of a lower level. Interaction of the models for calculation schemes of different levels is demonstrated in Fig. 2.

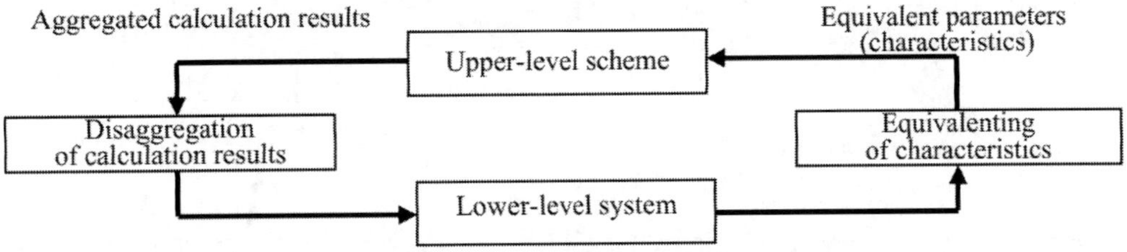

Fig.2. Interaction of models for calculation schemes of different levels.

3. Arrangement of information models

The multi-level modeling can be realized by applying the multi-level arrangement of data according to the following principles [5]: the modeled HSS can consist of sets of calculation schemes of different types; the scheme of a certain type is generated from its mix of elements; each element type can be preset in one of two formats: node, link; a real element of HSS can be represented by the aggregated element at the upper level, and by the own calculation scheme at the lower level; each element can be in one of several states (connected, disconnected, etc.); all elements of a certain type have the same set of parameters; the information model contains both the data on heat supply system elements, and the information on interrelations between: graphical representations of each element and its data, scheme elements, different schemes, each scheme and its layout, each scheme and its type of the problem solved, each problem and its calculation scheme. Each element has: number of the scheme it belongs to; number of the lower-level scheme, which is represented by it at its level; number of the upper-level scheme, with which the scheme including this element is associated. Application of such interrelations between the elements allows the modeling of diverse relations in the hierarchy of calculation schemes, i.e. both the tree-like and multiloop structure of relations between the levels.

4. Technology of development of HSS operation conditions

The purpose of operational conditions development is to find such parameters of control elements (heat sources, pumping stations, central heat points), the composition and parameters of the control devices on the network and the consumers. These parameters should ensure the supply of all consumers with the required amount of heat of a given quality, taking into account all technical and technological requirements. The technology of operating condition development is treated in detail in [16] and consists of four main stages.

1. Development of the computer model of HSS in accordance with the accepted structure of calculation scheme decomposition (see Fig.1), which consists in creation of the graphic database for HSS and includes graphical presentation of the calculation scheme on the urban development plan and input of information on all system elements into the database. An example of the multi-level model of HSS is given in Fig. 3.

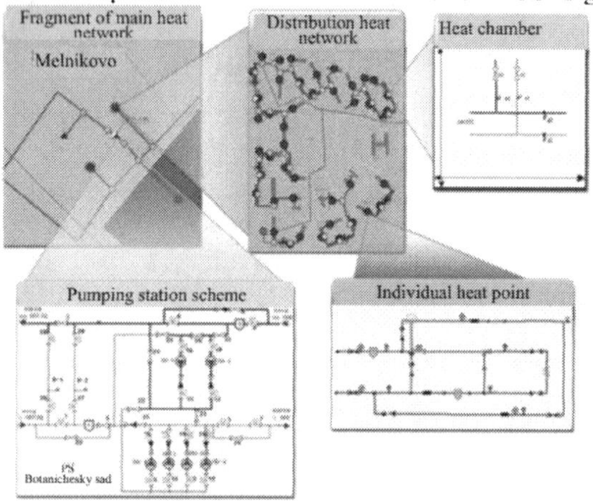

Fig. 3. Example of multi-level HSS model.

2. Analysis of the HSS conditions under the formed situation of operation to reveal bottlenecks, violations of the HSS condition and the quantitative estimation of the available level of heat supply to consumers. For this purpose the calculations of the thermal-hydraulic condition of HSS for the existing scheme of their operation and parameters at heat sources and central heat points which were maintained in the previous heating season. The calculation results are checked for condition feasibility in accordance with the requirements in [16].

3. Development of the feasible operation condition of HSS which will supply a required amount of heat of the specified quality for all consumers provided the requirements for operating parameters of all elements and minimum possible values of pressure in the network and heads in the pumping stations are observed. This stage includes a series of calculations of thermal-hydraulic conditions of HSS. In the case of rugged topography, which considerably narrows a feasible condition region, a multi-stage throttling is applied in the network and at consumers. The task to choose a place for throttler installation on the network is considered in [17, 18]. To develop the operating condition the parameters of control elements are chosen using the master nodes and consumers [16].

4. The multi-level adjustment calculation of thermal-hydraulic conditions by the approach in [19–22] with determination of individual corrections of heat carrier flow rates to compensate for heat losses in the networks for each consumer. If the network throughput is

insufficient, the temperature chart is corrected according to [23–24]. Then the parameters of throttlers at consumers with further verification of thermal-hydraulic conditions are calculated and the level of consumer supply is estimated. In the case of limited network throughput the analysis, whether the specific pressure losses are exceeded, is carried out and the pipeline sections to be re-laid for a larger diameter are selected. Sectionalizing valves, if available, can increase network throughput by changing the scheme of HSS sectioning [25].

5. Programs and their application

The methods for multi-level modeling of thermal-hydraulic conditions are realized in the ICS "ANGARA-TS [6, 12]. It is intended for automated development of operating conditions of HSS of an arbitrary structure (with any number and location of heat sources, pumping stations, heat points, consumers), arbitrary configuration (branched, multi-loop) and practically of any dimension. Herein, it is possible to calculate an individual level (for example, main or distribution network), and perform multi-level calculation of HSS with automatic coordination of boundary parameters between the levels. The ICS is applied to solve the following types of tasks.

Calculation tasks (single- and multi-level): adjustment and verification calculations of hydraulic condition; adjustment and verification calculations of thermal conditions; adjustment and verification calculations of thermal-hydraulic conditions; calculation of temperature charts of heat supply from heat sources; calculation of parameters of throttlers on the subscriber systems and in the heat network.

Analytical tasks: construction of the piezometric graphs; construction of temperature charts of heat supply from heat sources; determination of violations for state parameters, taking into account the limitations on admissible values of pressures, flow rates, speeds, temperatures and other parameters.

Let us consider the principle of multi-level modeling by the example of the real heat supply system in the Petropavlovsk-Kamchatsky city. The city is supplied with heat by 2 cogeneration plants and 30 boiler plants. The heat supply scheme is closed, independent with connection of distributed networks to the main networks through heat exchangers at the central heat points. The number of central heat points in the city totals 86. The problems of consumer provision with thermal energy are caused by: highly broken terrain; great heterogeneity of the schemes of consumer connection within a distribution network; misalignment of networks; excessive pump capacity at the central heat points; technical impossibility to drop excessive heads at the points of connection to buildings with low loads and direct connection of heating systems.

Development of conditions and adjustment measures on the basis of the suggested approach enabled to improve heat supply quality of the city and reveal and realize, in parallel, a high potential of energy saving owing to: considerable decrease of circulating water flow rate; decrease of make-up water consumption and water discharges by population; reduction of heat losses; decrease of electricity consumption for heat carrier pumping and chemical water treatment; switching of work of some central heat points with direct connection of heating systems to a lower temperature chart due to decease in flow rates. The economic effect of development of conditions and adjustment measures [26], carried out on the basis of calculations, amounted to 182 thousand rubles for one central heat point during the first year after their implementation. For the total number of central heat points in the city (86) the total expected economic effect will reach 15 million rubles annually.

6. Conclusion

The hierarchical structure of large HSS with intermediate control stages naturally involves the use of multi-level approach to modeling their conditions. Application of multi-level modeling enables to separate networks into fragments (main and distribution networks) even if there are no real central heat points among them, which essentially reduces problem dimension. The mathematical models and methods realized in the ICS "ANGARA-TS" are used to calculate networks within one fragment with a dimension of up to 32000 links. Hence, the existing large HSS practically of any dimension can be modeled with decomposition into levels.

A special methodology was developed for application of the multi-level approach to modeling of operation conditions of large HSS. It comprises the methods of multi-level adjustment thermal-hydraulic calculation of heating systems and hot water supply systems, determination of master consumers, the method of multi-step throttling on the network and by consumers.

Application of the multi-level modeling methods to develop operating conditions of real HSS in cities considerably improved their heat supply quality and revealed a high potential of energy saving.

References

1. A.V. Alekseev, N.N. Novitsky et al., *Pipeline systems in energy sector*, 211 (Novosibirsk: Nauka, 2008).
2. N.N. Novitsky, A.V. Alekseev et al., *Information and mathematical technologies in science and management*, 110 (Novosibirsk: Nauka, 2007).
3. A.V.Alekseev et al., *Pipeline systems in energy sector*, 221 (Novosibirsk: Nauka, 2007).
4. Z.I. Shalaginova, N.N. Novitsky, V.V. Tokarev, O.A. Grebneva, *Energy of Russia in the XXI century. Innovation development and management*, 389 (Irkutsk: ESI SB RAS, 2015).
5. A.V. Alekseev, N.N. Novitsk, Z.I. Shalaginiva, *Informstion and mathematical technologies in scientific research*, **XI**, 7 (Irkutsk ESI SB RAS, 2006).
6. N.N. Novitsky, V.V. Tokarev, Z.I.Shalagonova et al., Information and mathematical technologies in science, engineering and education, **X (II)**, 285 (Irkutsk, ESI SB RAS, 2005).
7. N.N. Novitsky, V.V. Tokarev, Z.I.Shalagonova et al., *Investigations and developments of Siberian Branch of Russian Academy of Sciences in the area of energy efficient technologies*, 48 (Novosubirsk: Nauka, 2009).
8. N.N. Novitsky, Thermal Engineering, **56 (12)**, 1010 (2009).
9. Z.I. Shalaginova, Thermal Engineering. **56 (12)**, 1016 (2009).
10. V.G. Sidler, Z.I. Shalaginova, Improvement of control and automation of heating systems, **VIII**, 37 (Kharkov, 1988).
11. Z.I. Shalaginova, Human-Environment-Universe, 182 (Irkutsk: IrSTU, 1997).
12. N.N. Novitsky, V.V. Tokarev, Z.I. Shalaginova, Information systems and technologies in energy sector and housing and public utilities, 50 (Ukraine, Kharkov: KHNURE, 2011).
13. Z.I.Shalaginova, Doctoral thesis in engineering, (Irkutsk: ESI SB RAS, 1995).
14. Z.I. Shalaginova, Mathematical models and methods for analysis and optimal synthesis of developing pipeline and hydraulic systems, **XII**, 315 (Irkutsk: ESI SB RAS, 2010).
15. Z.I. Shalaginova, Thermal Engineering, **63 (3)**, 222 (2016).
16. N.N. Novitsky, Z.I. Shalaginova, V.V. Tokarev, Izvestiya RAN: Energetika, **1**, 12 (2018).
17. A.V. Lutsenko, Systems studies in energy, **46**, 28 (Irkutsk: ESI SB RAS, 2016).
18. N.N.Novitsky, A.V. Lutsenko, Information and mathematical technologies in science and management, **XX**, 138 (Irkutsk: ESI SB RAS, 2015).
19. V.V. Tokarev, Z.I. Shalaginiva, mathematical models and methods for analysis and optimal synthesis of developing pipeline systems, **XII**, 300 (Irkutsk: ESI SB RAS, 2010).
20. N.N. Novitsky et al., *Pipeline systems in energy sector,* (Novosibirsk: Nauka, 2015).
21. V.V. Tokarev, Z.I. Shalaginova, Thermal Engineering, **63 (1)**, 68 (2016).
22. Z.I. Shalaginiva, V.V. Tokarev, O.A. Grebneva, Proceedings of IrSTU, **3 (98)**, 165 (2015).
23. Z.I Shalaginova, Thermal Engineering, **51 (7)**, 554 (2004).
24. Z.I. Shalaginova, Pipeline systems in energy sector, 372 (Novosibirsk: Nauka, 2004).
25. V. V. Tokarev, Thermal Engineering, **65 (6)**, 400 (2018).

PLANNING OF HEAT SUPPLY OF TERRITORIES: MODEL AND ALGORITHMS

Ekaterina E. Mednikova [1,*]

[1] Melentiev Energy Systems Institute of Siberian Branch of the Russian Academy of Sciences (ESI SB RAS), Pipeline Energy Systems Department, 130, Lermontov str., Irkutsk, Russia, 664033

Abstract. The choice of direction of heat supply systems development in cities is an important task for the authorities, district heating companies and consumers. The author focuses on two main problems in heat supply development questions: territory zoning by type of heat supply and determination of economically feasible heat supply systems sizes. For the solution of these problems, the mathematical model and algorithms were developed. For the solution of this task algorithms were developed.

1 Introduction

District heating (DH) is wide spread in Russia, because there are special climatic and economic conditions for its development and implementation in this country. DH covers more than 70% of heat energy consumption. It is necessary for all settlements where DH systems operate to develop and approve the scheme of heat supply to ensure efficient and reliable development of heat supply system(HSS).

Last years, the requirements to quality and reliability of heat supply were changed. The regulatory documents for Russian heat supply systems were adopted, such as the Russian Law "On heat supply" [1] and "Requirements to heat supply schemes" [2]. The analysis of regulatory and legal framework shows that problems of energy planning are important. Many reasons led to necessity to review approaches for planning of heat supply development [3]. Considering the above, it's necessity to make complex assessment of all competitive variants to choose the type and capacity of DH system, to increase its efficiency and reliability. In this case, several problems must be solved, these are task of territory zoning by type of heat supply, justification of optimal size of HSS and concentration of heat sources capacities. The transition to innovative solutions in the field of heat supply [4, 5], creation of the intelligent HSS are impossible without effective planning of HSS development and taking reliability [6] and consumers requirements into consideration.

In Russia, heat supply planning is not so developed like in Europe, especially in the Scandinavian countries [2-4, 7-14]. The basis of heat supply planning are presented in works [7-9, 15]. Besides, the importance of aforementioned problems is mentioned in other countries [16, 17]. The use of existing European approaches doesn't guarantee a reliable solution for Russian HSSs for many reasons, for example, different technical solutions and

[*] Corresponding author: yakimetse@isem.irk.ru

characteristics of heating system; different climatic conditions and specifications and technical documentation [18].

This work continues and develops the basic provisions of previous works [3, 19] and it's based on the Theory of Hydraulic Circuits [20]. The author proposed several algorithms to carry out the territory zoning by type of heat supply and to justify economically feasible heat supply systems' sizes.

2 Heat supply planning model

2.1 Heat supply planning objective

Proposed heat supply planning technique allows to find optimal solution for two main problems. These are territory zoning by type of heat supply and search of economically feasible size of a DH system. Three mathematic models were worked out. The first is for complex heat supply development planning and two others [3] are for solving the problems individually. This paper presents the complex technique.

The objective of the heat supply planning model is to minimize the total cost of heat supply including generation cost, transmission and distribution costs.

$$C^{HSS}(Q, x, t, q_S, q_L) = C^{DHS}(Q, x, t, q_S, q_L) + C^{IND}(Q, t) \rightarrow \min \qquad t \in T \quad (1)$$

where C^{HSS} – total cost of heat supply (EUR), C^{DHS} – total cost of DH (EUR), C^{IND} – total cost of individual heating (EUR), Q – heat load (MW), x – a heat carrier flow at the sections of heat supply system's network (t/h), q_S is the boundary value of heat load density (MW/km2), q_L is the boundary value of linear heat density (GJ/m), t is an index for years in the scheduling horizon.

The objective includes the annual investment planning and operation costs associated with generating units (CHP, heat-only boilers, and individual heat units) and heat pipelines for systems with an equal energy effect. Total cost of DH includes the cost of heat production, heat transmission and distribution and additional specific investment cost for increasing of heat source capacity in DH system [21]. Total cost of individual heating includes fuel costs and additional specific investment cost for the unit of individual heating.

The heat supply planning model includes the system of constraints and conditions [3]: energy balance between heat production and heat consumption, capacity constraints for heat sources in DH systems and individual units, equations for calculation of the flow distribution in hydraulic circuit, equation that describes correlation between a mass flow rate and a heat load, conditions imposed by requirements of heat supply reliability [19], which are defined by standard values of nodal indices of reliability, equations for estimation of boundary values of heat density criteria q_S and q_L, equation for estimation of actual value of heat load density per unit area, equation for estimation of an actual value of linear heat density (for transmission pipes).

Condition of district's connection to DH system (2):

$$\begin{cases} \text{If } HD \geq q_s \text{ for the district, DH for this district is considered economically feasible,} \\ \text{If } HD < q_s \text{ for the district, individual heating for this district is considered economically feasible.} \end{cases}$$

where HD is an actual value of heat load density for selected district (MW/km^2).

The actual value of linear heat density should be higher than the normative value:

$$LHD_i \geq q_{Li} \quad (3)$$

where LHD_i – an actual value of linear heat density for i-th heat source (GJ/m), i is an index for heat sources in DH zone.

Nonlinear programming methods are used to solve the optimization problem.

If the problems need to be solved separately, their objective functions can be expressed according to the Ref.[3]:

objective function for territory zoning by type of heat supply

$$C^{HSS}(Q,x,t,q_S) = \sum_i C_{S_i}^{DHS}(Q,t) + \sum_m C_{HN_m}(x,t,q_S) + \sum_n C_n^{IND}(Q,t) \rightarrow min \quad (4)$$

$$i \in I, m \in M, n \in N, t \in T$$

objective function for determination of economically feasible size (scale) of DH system

$$C^{HSS}(Q,\ x,t,\ q_L) = \sum_i C_{S_i}^{DHS}(Q,t) + \sum_m C_{HN_m}(x,t,\ q_L) \rightarrow min$$

$$i \in I,\ m \in M,\ t \in T \quad (5)$$

where $C_{S_i}^{DHS}$, C_{HN_m}, C_n^{IND} are total costs for heat sources in DH zone, heat network in DH system and individual heating units (EUR).

2.2 Heat supply planning algorithms

The complex technique for heat supply planning of cities territory has been divided into two algorithms for usability like it's presented in Fig.1.

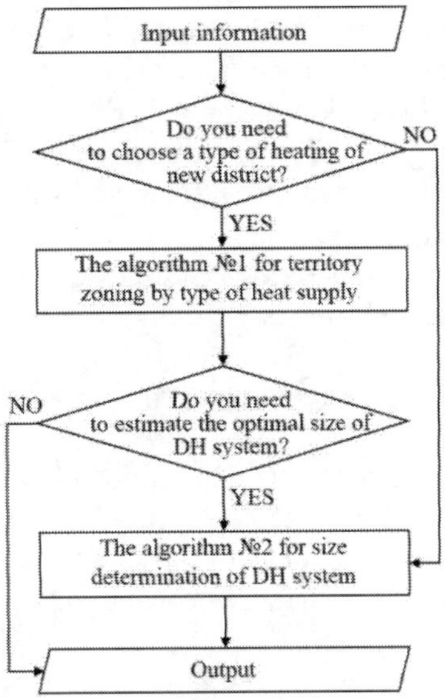

Fig. 1. Sequence of the solution of HSSs development problems

The algorithm №1 includes the following steps:

1. Organization of Data base, including territorial planning, buildings types, expected heat loads of consumers, etc.
2. Development of forward variants of the territory division according to the Urban plan. This step means the selection of districts of perspective development where it is necessary to organize heat supply.
3. Calculation of these districts' area.

4. Calculation of consumers' heat loads in the perspective districts.
5. The choice of comparable sources (for DH system and individual heating) and input technical and economic information of these heat sources.
6. Calculation of the actual value of heat load density per area unit for selected districts.
7. Calculation of the boundary value of heat load density per area unit q_S for comparable heat sources, Ref. [3].
8. Estimation of collected actual values of heat load density with regard to the boundary value q_S and zoning territory by types of heat supply.

8.1. If actual value of heat load density for selected district is higher than the boundary value, the considered district belongs to the DH zone.

8.2. If actual value of heat load density for selected district is lower than the boundary value, the considered district belongs to the individual heating zone.

9. Check of feasibility of heat balance.
10. Check of capacity constraints for heat sources in DH systems and individual units.
11. This step depends on the problem statement.

11.1. Calculation of the objective function (4) if the problem of territory zoning by type of heat supply needs to be solved separately.

11.2. The transition to the implementation of the algorithm №2 if the problem needs to be solved comprehensively.

The second algorithm is the algorithm for size determination of DH system. It has an objective function (5) if the problem needs to be solved separately.

The algorithm №2 includes the following steps:

1. Organization of Data base for DH zones, including existing network topology, node and pipe variables, etc.
2. Choosing of heat sources and its location. Here you can choose existing CHP plants and heat-only boilers or new heat sources to cover the power deficit.
3. Calculation of heat loads for each considered heat sources. At this stage, it is necessary to take into account all consumers on considered period.
4. Calculation of the flow distribution in hydraulic circuit (hydraulic calculation).
5. Check of feasibility of heat balance.
6. Check of capacity constraints for heat sources in DH systems.
7. Calculation of transmission pipes' length and the actual value of linear heat density for considered heat sources.
8. Calculation of the boundary value of linear heat density q_L for considered heat sources, Ref. [3].
9. Comparison of collected actual values of linear heat density with boundary values q_L and preliminary evaluation of DH systems sizes.

9.1. If linear heat density for selected heat source and connected to it consumers is higher than the boundary value, considered consumers belong to the operation zone of this heat source (go to the step 10).

9.2. If linear heat density for selected heat source and connected to it consumers is lower than the boundary value, this DH system should be divided into several small DH systems for increasing of actual linear heat density up to boundary value. This can be done by consecutive disconnection of the most removed consumers or by redistribution of loads among sources. Return to the step 2 if you need to choose another heat source or to the step 3 if you don't need to choose another heat source.

10. Check of conditions imposed by requirements of heat supply reliability, which are defined by standard values of nodal indices of reliability.

10.1. In case of violation of the constraints, duplication of pipeline sections and connectors between pipelines must be added. Also you can increase reliability by redistribution of loads among sources. Return to the step 2.

10.2. Go to the next step if requirements are satisfied.

11. This step depends on the problem statement.

11.1. Calculation of the objective function (5) if the problem of size determination of DH system needs to be solved separately.

11.2. Calculation of the objective function (1) if the problem needs to be solved comprehensively.

3 Case study

The case study example presented in this section illustrate application of the proposed heat supply planning model for Shelehov city with population over 47 thousand (Irkutsk region, Russia), see Fig. 2. HSS includes one CHP (heat power – 404 MW) with transmission and distribution heat pipelines and 1383 dwelling houses with individual heat installations with total heat load about 60 MW. The length of transmission pipes is 15.7 km.

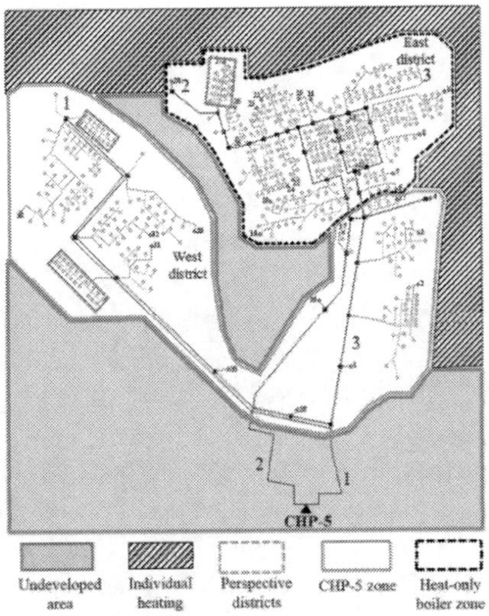

Fig. 2. The heat supply scheme of Shelehov city.

The case study results are compared and analyzed based on the objective function (1) because we need to solve both problems of DH system's development and take into account the reliability requirements. Territory zoning by type of heat supply is carried out based on the algorithm №1. Fig. 2 shows zones of district and individual heating. Individual heating zone is districts with heat load density less than 22 MW/km^2. Analysis of DH system's economic feasibility is based on the algorithm №2.

The findings of the analysis indicate that DH zone should be divided into two systems with different heat sources. The first system is the West district and part of the East district and existing CHP, the second system is a new heat-only boiler and part of consumers of the East district. According to the results, new consumers located within the received zones have to be connected to the relevant systems. According to preliminary estimates, the organization of heat supply under the new scheme taking into account the allowable reduction in the heat production for CHP-5 will reduce the total costs by 15%, and the specific costs of heat energy distribution will be reduced by 14%.

4 Conclusion

The mathematical model and algorithms oriented to modern conditions of HSSs development are offered for the solution of heat supply planning problems. The importance of the offered mathematical statement is the possibility to use it at the stage of predesign researches for preliminary assessment of directions of heat supply systems development. The offered mathematical model includes restrictions for heat load density, which allow to reduce the number of variables in the calculated schemes. On the one hand, the findings in this paper allow to plan development of the heat supply systems reasonably. On the other hand, the results create good conditions for more detailed calculations for the choice of optimal composition of heat sources, their capacities and types, and also a configuration of the heat network and parameters of its elements.

Acknowledgements

The research was performed at Melentiev Energy Systems Institute SB RAS in the framework of the scientific project III.17.4.3 №AAAA-A17-117030310437-4 of the foundation researches program of SB RAS.

References

1. Federal Law of the Russian Federation N 190-FZ of 27.07.2010 «On heat supply»
2. The Governmental order № 154 of 22.02.2012 «On requirements to heat supply schemes, the order of their development and approval».
3. V.A. Stennikov, E.E. Iakimetc, Energy, **110**, 157-165 (2016)
4. H. Lund, S. Werner, R. Wiltshire, S. Svendsen, J.E. Thorsen, F. Hvelplund, B.V. Mathiesen, Energy, **68**, 1-11 (2014)
5. J. Ziemele, A. Gravelsins, A. Blumberga, D. Blumberga, Energy Proceedia, **95**, 522-527 (2016)
6. V.A. Stennikov, I.V. Postnikov, In: Vasant P., Voropai N.I., editors. *Sustaining power resources through energy optimization and engineering* (Hershey PA: Engineering Science Reference (an imprint of IGI Global)), 102-126 (2016)
7. IEA. District heating distribution in areas with low heat demand density. SenterNovem (2008)
8. K. Sperling, F. Hvelplund, B.V. Mathiesen, J. Energy Policy, **39**, 1338-1351 (2011)
9. U. Persson, S. Werner, J Applied Energy, **88**, 568-576 (2011)
10. K. Hansen, D. Connolly, H. Lund, D. Drysdale, J.Z. Thellufsen, Energy, **115**, 1663-1671 (2016)
11. B. Möller, H. Lund, Applied Energy, **87**, 1846-1857 (2010)
12. U. Persson, S. Werner, Applied Energy, **88**, 568-576 (2011)
13. S. Nielsen, B. Möller, Energy, **57**, 458–468 (2013)
14. L. Grundahl, S. Nielsen, H. Lund, B. Möller, Energy, **115**, 1771-1778 (2016)
15. S. Frederiksen, S. Werner, *District heating and cooling.* (Studentlitteratur AB, 2013)
16. Z. Huang, H. Yu, Energy and Buildings, **71(7)**, 103-111 (2014)
17. G. Mendes, C. Ioakimidis, P. Ferrao, Ren. and Sust. Energy Rev., **15**, 4836-4854 (2011)
18. A. Nuorkivi, Advanced District Heating and Cooling (DHC) Systems. (Woodhead Publishing is an imprint of Elsevier, 2016), 17-41
19. V.A. Stennikov, I.V. Postnikov, J Power Technology and Engineering, **47(6)**, 446-453 (2014)
20. A.P. Merenkov, V. Ya. Khasilev, *The Theory of hydraulic circuits,* (Moscow, Nauka,. 1985) (in Russian)
21. V.A. Stennikov, E.E. Mednikova, Thermal Engineering, **63(9)**, 657-665 (2016)

Optimization of reliability parameters structure for district heating systems

Ivan V. Postnikov[1,*]

[1]Melentiev Energy Systems Institute of Siberian Branch of the Russian Academy of Sciences (ESI SB RAS), Pipeline Energy Systems Department, 130, Lermontov str., Irkutsk, Russia, 664033

Abstract. The problem of reliability parameters distribution on the district heating system's scheme is considered. This is one of important part of the general problem of optimal synthesis of district heating systems and is urgent for both the systems under design and the existing insufficiently reliable systems. The concept of solving the problem is based on the average reliability parameters of components (failure and restoration rates) which determine the first approximation to the optimal solution. This parameters needs for the further distribution its average values among system components. Algorithm and mathematical models for determine of optimal reliability parameters of system components with provide the required level of heat supply reliability and minimal total costs on ensuring this level are developed. The methodology of solving the stated problem is based on the methods of the theory of hydraulic circuits, nodal reliability indices, models of Markov random process and general regularities of heat transfer processes. The methodology also takes into account changes in thermal loads during the heating period and time redundancy of consumers related to heat storage.

Keywords: district heating system; reliability optimization; nodal reliability indices; component reliability; failure and restoration rates; Markov random process.

1 Introduction

District heating systems (DHS) is the most important component in support of vital activity of population and development of all economic branches. High socio-economic significance of the heat supply sphere imposes heavy demands to reliability of DHS that combine heat supply sources (HS) and heat networks (HN) in the unified structure.

There exist different methods for the analysis and optimization of reliability of HS and energy sources in general. These methods can be divided into analytical ones [1–5] based mainly on the Markov or semi-Markov processes, logical-and-probabilistic methods and methods of statistical modeling [1, 6, 7].

The general principles of calculating reliability and redundancy of HN of large DHS were formulated in 1972 in [8]. The approach that is based on the evaluation of nodal reliability indices was developed later at Siberian Energy Institute, now it is Melentiev

* Corresponding author: postnikov@isem.irk.ru

Energy Systems Institute of SB RAS [1, 9]. Along with the nodal approach to reliability assessment of HN there is another concept that is based on the integral reliability index [10]. Many aspects of HN reliability problems are common with similar tasks for other energy networks or pipeline systems (oil, gas, electricity etc.), e.g. [11–15].

It should be noted that all these methods relate to separate subsystems of DHS – HS or HN. This don't let to get a total system estimation of heat supply reliability. For deliverance of this shortcoming the authors worked out several methods and models for comprehensive (complex) analysis of DHS, e.g. [16–19]. This paper is devoted to optimization of component reliability as one of the major tasks of reliability synthesis of DHS.

2 Methodology for solving the problem of reliability parameters structure optimization for DHS

The specified values of reliability parameters of DHS components are improved or provided by reducing their *failure rates* and improving their *restoration rates* that is reduction in the restoration time. Statement of the optimization problem of reliability parameters structure optimization for DHS consists in determining the reliability parameters of it components (failure and restoration rates), which provide the required level of heat supply reliability at the minimum costs of ensuring these parameters and constraints on technically possible values. According to the methodology for DHS reliability analysis which is presented in [1], the required level of heat supply reliability is determined by the standard values of two nodal reliability indices: failure-free operation probability (FOP) R_j^0 and availability factor (AF) K_j^0, that are specified for all $j \in J$, where j – number of consumer, J – set of consumers.

Solving of the stated problem based on the concept of *average reliability parameter* of components. Average reliability parameter of components is taken to mean their failure or restoration rate that preliminarily has the same value for these components, which provides the required level of reliability indices. These parameters are determined by the formulas intended for the calculation of nodal reliability indices of DHS [1, 16, 19], Rossander equation that determines annual heat load curves of consumers [20], and some basic laws of district heating and thermal physical processes involved in DHS [20].

Thus, mathematical formulation for the optimization of reliability parameters of DHS components is follow:

$$F_{\text{obj}} = \sum_{n\in N} f_{n\lambda}(\lambda_n) + \sum_{n\in N} f_{n\mu}(\mu_n) \to \min; \tag{1}$$

$$\bar{\lambda}_j = \frac{1}{\tau_0}[\ln(1/R_{oj})] \times \left(1 - N_s(1-K_{oj}) \Big/ \sum_{s\in E} L_s^{\sigma_j}\right)^{-1} \times \left(\sum_{s\in E} M_s^{\sigma_j}\right)^{-1}; \tag{2}$$

$$L_s = \frac{1}{1-\omega_j}\left[1 - \frac{1}{q_{oj}}\left(q_{sj} + \varphi_j t_{sj} - \varphi_j \frac{C_1 - C_2 \exp B_j}{C_3(1-\exp B_j)}\right)\right]; \tag{3}$$

$$M_s = \frac{1}{1-\omega_j}\left(1 - \bar{q}_{sj} + \frac{\varphi_j t_{sj}}{q_{oj}} - \varphi_j \frac{C_1 - C_2 \exp B_j}{C_3 q_{oj}(1-\exp B_j)}\right); \tag{4}$$

$$\bar{q}_{sj} = q_{oj} / q_{sj}, s \in E, j \in J; \tag{5}$$

$$C_1 = t_{oj}(1-\bar{q}_{sj}), C_2 = t_{j\min} - t_{oj}\bar{q}_{sj}, C_3 = 1 - \bar{q}_{sj};\ B_j = 1/(\varepsilon_j \bar{\mu}_j),\ j \in J; \tag{6}$$

$$\bar{\lambda}_j \sum_{s\in E} p_s = \sum_{n\in N} \sum_{s\in E(n)} \lambda_n p_s,\ j\in J; \tag{7}$$

$$\bar{\mu}_j \sum_{s\in E} p_s = \sum_{n\in N} \sum_{s\in E(n)} \mu_n p_s,\ j\in J; \tag{8}$$

$$p_s\left(\sum_{n\in N(s)} \lambda_n + \sum_{n\in N(s)} \mu_n\right) = \sum_{z\in E(s)}\left(\sum_{n\in N(z)} p_z\lambda_n + \sum_{n\in N(z)} p_z\mu_n\right), s\in E; \tag{9}$$

$$\mathbf{A}\mathbf{x}_s = \mathbf{q}_s,\ s\in S; \tag{10}$$

$$\overline{\mathbf{A}}_s^{\mathrm{T}}\mathbf{p}_s = \mathbf{h}_s - \mathbf{H}_s,\ s\in S; \tag{11}$$

$$\mathbf{S}\mathbf{X}_s\mathbf{x}_s = \mathbf{h}_s,\ s\in S. \tag{12}$$

$$\lambda_n^{\min} \le \lambda_n \le \lambda_n^{\max},\ n\in N;\ \ \mu_n^{\min} \le \mu_n \le \mu_n^{\max},\ n\in N; \tag{13}$$

Here: $f_{n\lambda}(\lambda_n)$, $f_{n\mu}(\mu_n)$ – cost functions of ensuring reliability parameters of components, i.e. their failure and restoration rates, respectively, rub.; $\bar{\lambda}_j$ и $\bar{\mu}_j$ – average failure and restoration rates for consumer j, respectively, 1/h; τ_0 – time instant corresponding to a total number of hours of the considered (heating) period, h; N_s – the quantity of system states; s – number of system state; E – set of system states; ω_j, σ_j – irregularity factors of heat load curve of consumer j [20]; φ_j – coefficient of specific heat losses for consumer j, GJ/(h°C); t_{sj} – current (actual) internal air temperature for consumer j in system state s, °C; $\bar{q}_{sj}$ – relative heat supply to consumer j in system state s, GJ/h; t_{0j} – design temperature of internal air for consumer j, °C; $t_{j\min}$ – minimum admissible temperature of internal air for consumer j, °C; q_{sj} – level of heat supply to consumer j in system state s, GJ/h; ε_j – coefficient of thermal energy storage for consumer j, h; $L_s, M_s, C_1, C_2, C_3, B_j$ – assumed abbreviations of expressions; p_s – probability of system state s; n – number of system component; N – set of system components; $E(n)$ – is a subset of system states for which the system can transition because of failure or restoration of component n; λ_n, μ_n – failure or restoration rates of component n, 1/h; p_z – probability of the system state z (division of state into s and z is necessary to write the system of equations of random process); $N(s)$ – subset of system components whose failure or restoration corresponds to a direct transition of the system from state s to some other state z; $N(z)$ – subset of system components whose failure or repair corresponds to a direct transition of the systems from state z to some other state s; $E(s)$ – subset of the system states from which the system can transition to state s; $\mathbf{A}_s$ – incidence matrix of linearly independent nodes in the network under emergency system state s (considering failure of some component); $\overline{\mathbf{A}}_s^{\mathrm{T}}$ – full transposed node-branch incidence matrix; $\mathbf{x}_s$ – vector of heat carrier flow rates in the network sections (branches) under emergency system state s, t/h; $\mathbf{q}_s$ – vector of flow rates at network nodes under emergency system state s, t/h; $\mathbf{p}_s$ – vector of nodal pressures of HN under emergency state s, mm wc; $\mathbf{h}_s$ – vector of heat losses in the sections under the emergency system state s, mm wc; $\mathbf{H}_s$ –

vector of operating heads at sources in the emergency system state s, mm wc; $\mathbf{S}, \mathbf{X}_s$ – diagonal matrices of coefficients of hydraulic resistance of sections, m/(h2t2), made up from the values of hydraulic resistances of sections and absolute values of flow rates in them, t/h; $\lambda_n^{\min}, \lambda_n^{\max}$ – minimal and maximal available values of failure rate for component n (1/h); $\mu_n^{\min}, \mu_n^{\max}$ – minimal and maximal available values of restoration rate for component n (1/h).

The form of functions (1) and their quantitative parameters are determined by the methods of approximation on the basis of an analysis of the actual data mainly on the costs of equipment with different reliability characteristics and establishing and maintaining the emergency and restoration services. The rationale of using the Markov model (9) as well as the other aspects of using the tool of Markov random processes in the problems of DHS reliability are considered in more detail in [1, 16, 19]. The hydraulic conditions in HN are calculated by the methods of theory of hydraulic circuits [9] with using the nodal and matrix form of model of flow distribution (hydraulic conditions) in the network (10)–(12).

3 Case study

Consideration is given to DHS scheme presented in Fig. 1-a. The scheme consists of two district heat sources (HS1 and HS2), seven consumers (nodes 1–7) and circuit HN consisting of 18 sections (components). An aggregate scheme of components for both HS is demonstrated in Fig. 1-b,c and consists of boilers 19, 25; turbines 20, 26; network heaters 21, 23, 27, 29; network pumps 22, 24 and 28, 30.

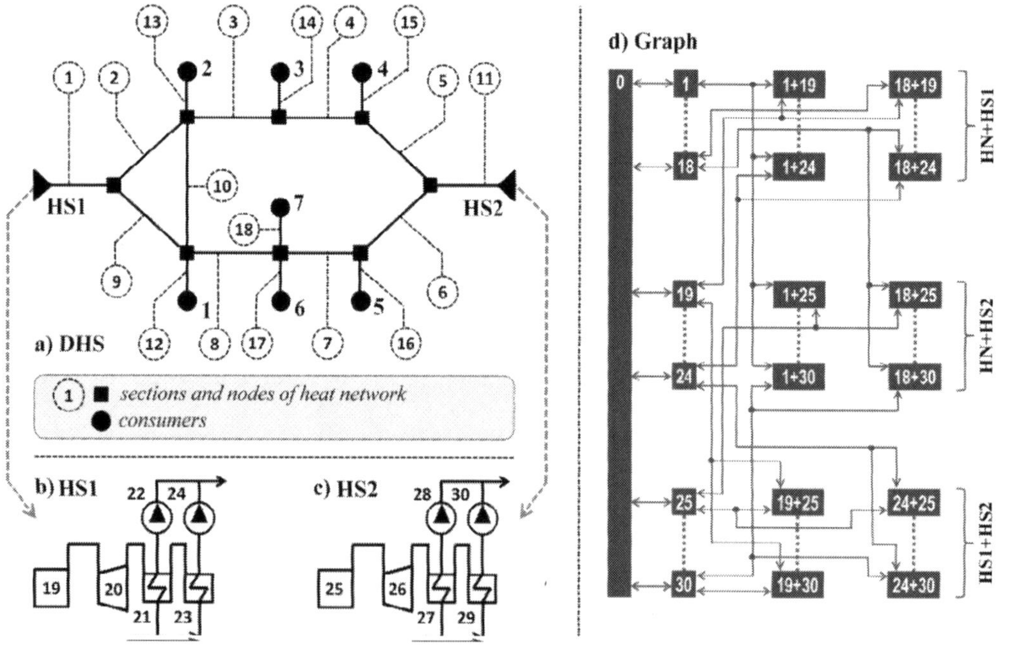

Fig. 1. Schemes of case study: a) general scheme of DHS; b) simplified scheme of HS1; c) simplified scheme of HS2; d) graph of DHS states and relation between them.

The random process of DHS operation is modeled for the following conditions: each component can be in two states – operable and failed, and the flow of events within one subsystem (HN, HS1 and HS2) is the simplest. The latter condition suggests a simultaneous failure of components only from different subsystems of DHS. Here we will confine ourselves to the consideration of the state when no more than 2 components fail at once. Thus, the set of states is formed by individual states of HN, HS1, HS2 and combinations of these states for HN+HS1, HN+HS2 and HS1+HS2. The graph of DHS states in a reduced form is presented in Fig. 1-d. The results of the calculation of probabilities of system states and post-emergency hydraulic conditions are not presented due to a large data array.

The reliability parameters of the DHS are optimized provided the following standard values of nodal reliability parameters [1] are met: for AF – 0,97 and FOP – 0,905. The following ranges of possible values of optimized reliability parameters of the DHS components are assumed: 0,0002–0,0025 1/h – for failure rate; 0,007–0,09 1/h – for restoration rate.

Results of search for an optimal solution on the choice of reliability parameters of DHS components with the required reliability level are presented on Fig. 2. The GAMS software package is used as a solver.

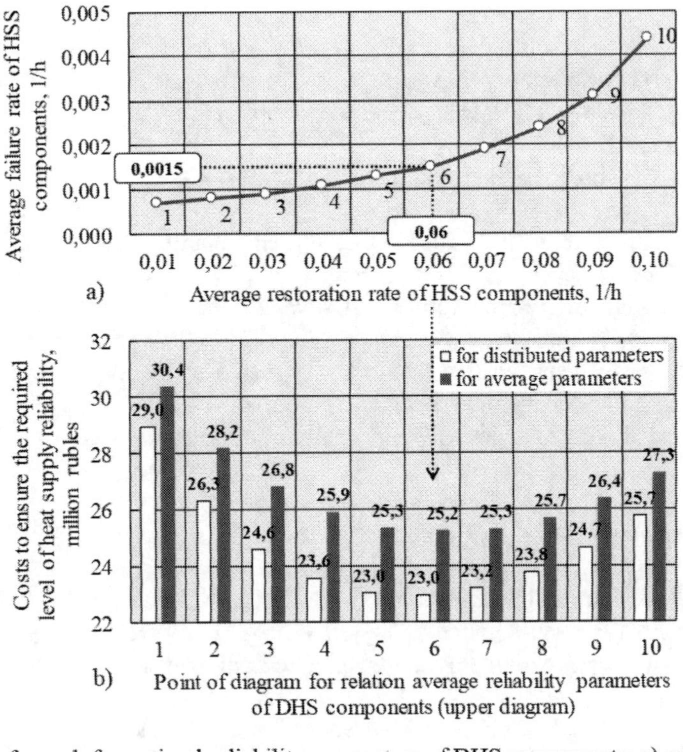

Fig. 2. Results of search for optimal reliability parameters of DHS components: a) relationship among average failure and restoration rates of DHS components; b) costs to ensure the required reliability level of heat supply to consumers.

The diagram on Fig. 2-a demonstrates the relationship between the average reliability parameters (failure and restoration rates) of the components. The diagrams on Fig. 2-b shows a change in the costs to ensure the required level of heat supply reliability depending on relationship of average reliability parameters for their optimal distributed among components and without this distribution. The solution obtained under the specified conditions corresponds to the reliability costs in the amount of 25,2 million rubles (per year) for average reliability parameters of components under the following values of average reliability parameters: failure rate – 0,0015 1/h, restoration rate – 0,06 1/h (point 6); and 23 million rubles for optimal distributed its average values among components. Thus, the economic effect with the optimal distribution of average reliability parameters among components is 2,3 million rubles or 9%.

4 Conclusions

The research suggests a methodology for the optimization of component reliability of DHS within the general problem of its reliability synthesis. The advantages of the proposed methodology compared to the existing approaches to solving this problem consist in joint optimization of the component reliability of HS and HN schemes, integration of procedures for the reduction in failure rates and the improvement in restoration rates of the components in the search of the optimal system reliability.

The research was performed at Melentiev Energy Systems Institute SB RAS in the framework of a scientific projects III.17.4.1 №AAAA-A17-117030310432-9 and III.17.4.3 №AAAA-A17-117030310437-4 of the foundation researches program of SB RAS.

References

1. E.V. Sennova, A.V. Smirnov, A.A. Ionin et al. *Reliability of heat supply systems* (Nauka, Novosibirsk, 2000) [In Russian]
2. M. Haghifam, M. Manbachi. International journal of electrical power & energy systems, **33**, 385–393 (2011)
3. A. Lisnianski, D. Elmakias, B.H. Hanoch. Reliability engineering & system safety, **98**, 1–6 (2012)
4. Gengfeng Li, Zhaohong Bie, Yu Kou, Jiangfeng Jiang, Mattia Bettinelli. Applied Energy, **167**, 397–406 (2016)
5. Hyung-Geun Kwag, Jin-O Kim. Applied Energy, **122**, 24–33 (2014)
6. A. Naess, B. Leira, O. Batsevych. Structural Safety, **31**, 349–355 (2009)
7. Liwei Shua, Lingen Chena, Jiashan Jina et al. Applied Energy, **80(1)**, 61–66 (2005)
8. V.Ya. Khasilev, M.K. Takaishvili. Teploenergetika, **4**, 14–19 (1972) [In Russian]
9. A.P. Merenkov, V.Ya. Khasilev. *Theory of hydraulic circuits* (Nauka, Moscow, 1985) [In Russian]
10. A.A. Ionin. *Reliability of heat network systems* (Strojizdat, Moscow, 1989) [In Russian]
11. Sigitas Rimkevicius, Algirdas Kaliatka, Mindaugas Valincius et al. Applied Energy, **94**, 22–33 (2012)
12. B.H. Zhou, Z.Q. Zhai. Engineering Failure Analysis, **18**, 1333–1340 (2011)
13. T. Adefarati, R.C. Bansal. Applied Energy, **185**, 158–171 (2017)
14. Nima Nikmehr, Sajad Najafi Ravadanegh. Electrical Power and Energy Systems, **78**, 80–87 (2016)
15. Ferdinando Salata, Andrea de Lieto Vollaro, Roberto de LietoVollaro et al. Energy and Buildings, **86**, 118–136 (2015)
16. V.A. Stennikov, I.V. Postnikov. Power Technology and Engineering, **47(6)**, 446–453 (2014)
17. I.V. Postnikov, V.A. Stennikov, E.E. Mednikova, A.V. Penkovskii. Applied Energy (article in press, 2017) DOI: doi.org/10.1016/j.apenergy.2017.11.073
18. I.V. Postnikov, V.A. Stennikov, E.E. Mednikova, A.V. Penkovskii. Energy Procedia, **105**, 3083–3088 (2017)
19. V.A. Stennikov, I.V. Postnikov. In: *Sustaining power resources through energy optimization and engineering*, editors P. Vasant, N.I. Voropai (Engineering Science Reference, Hershey PA, 2016)
20. V.Y. Sokolov. *Cogeneration and heat networks*. (Publishing House of MEI, Moscow, 1999) [In Russian]

Search for an equilibrium between heat demand and supply in heating systems under a unified heat supply organization

Andrey V. Penkovsky[1,*]

[1]Melentiev Energy Systems Institute of Siberian Branch of the Russian Academy of Sciences (ESI SB RAS), Pipeline Energy Systems Department, 130, Lermontov str., Irkutsk, Russia, 664033

Abstract. Organizational model of the heat energy market on a unified heat supply organization is considered. An equilibrium mathematical model is proposed for the liberalized tariff setting conditions based on a microeconomic monopolistic market model. This mathematical model allows for taking into account energy production and transportation costs as part of a single economic criterion and to determine the supply and demand equilibrium for heat energy. The search for an optimal solution to the developed mathematical model of the unified heat supply organization is based on the univariate relaxation method with the subsequent use of methods of redundant design and simple iterative schemes. The developed mathematical model fully reflect the current "rules of the game" between heat energy producers and consumers and allow a maximum consideration for the interests of all participants of the heat supply process under physical and engineering constraints on heat energy sources and heat networks.

1 Introduction

There are more than 50 thousand heat markets in the Russian Federation [1] and all of them are controlled by natural monopolies. Most of them represent local heating systems that have branched heat networks and a limited number of heat sources.

Currently, Russia's heat power industry goes toward liberalization and market reforms [2], which suggests abolishment of state control over heat tariffs and establishment of a unified heat supply organization (UHSO) according to the Federal law No.190 «On heat supply» [3]. This unified heat supply organization will make it possible to reduce the heat production costs and to more effectively supply heat to consumers in its service area.

This model of heat supply is being gradually adopted in real practice in urban and rural areas of the Russian Federation. The proposed model suggests the integration of all heating system functions, including heat generation, transportation and sale, to be performed within the unified heat supply organization.

With this heat supply model, the unified heat supply organization should become the owner of all the municipal assets including heat sources and distribution heat networks.

* Corresponding author: penkoffsky@isem.irk.ru

Such a merging of the main assets and heat supply control processes shapes the unified heat supply organization as a single seller in the heat market, i.e. a monopolist. Thus, the unified heat supply organization will have a total control of heat supply in the market and market heat price.

The relationships between the market participants in the form of the model of the unified heat supply organization develop according to a certain pattern and imply the following. Based on the forecasts of demand and optimal options of heating system expansion, the unified heat supply organization delivers heat (under medium-term and long-term contracts) to consumers at a price calculated as a sum of a price of heat source heat generation and a price of heat transportation from the heat source to the consumer. In this case, the unified heat supply organization produces the amount of heat that on the one hand would maximize its profit, given physical-technical constraints on heat sources and heat networks, and meet the demand for heat specified by the consumer, and on the other hand would correspond to the consumer wish to pay for this demand.

It is worthwhile to emphasize a classical microeconomic model of monopoly [4-7] among the most widely applied approaches to modeling of the medium–term (or long-term) forecasting of possible

2 A mathematical model of the Unified heat supply organization

Heat market in the format of an organizational model of the unified heat supply organization with a great number of sources and heat networks is modeled by a hydraulic circuit. The hydraulic circuit represents a calculated scheme of a real heating system that consists *of* m nodes and n branches [8]. The hydraulic circuit structure is described by a total coupling matrix $\overline{A}$. The number of rows in this matrix coincides with the number of nodes, and the number of columns – with the number of branches. Hydraulic circuit represents an aggregate of ordered sets, including a set of nodes – $J = \{j : j = 1,...,m\}$ that consists of subsets: J_{hs} – heat source, J_{con} – consumers and J_0 – ordinary branching nodes in the scheme; and a set of branches – $I = \{i : i = 1,...,n\}$, that show the specified pair-wise relations between the nodes.

Modeling of such a system is performed with a certain time interval, that starts with the initial time τ_0 (for example, corresponding to a calculated heating load) and finishes with the final (design) time T.

According to [9], the optimal flow distribution in the heat network for each calculated time τ can have the following form:

$$\mathrm{A}\mathrm{x}_\tau = \mathrm{G}_\tau, \quad (1)$$

$$\overline{\mathrm{A}}^{\mathrm{T}}\overline{\mathrm{P}_\tau} = \mathrm{h}_\tau - \mathrm{H}_\tau, \quad (2)$$

$$\mathrm{h}_\tau = \mathrm{S}\mathrm{X}_\tau \mathrm{x}_\tau, \quad (3)$$

where A – an $(m-1)\times n$ coupling matrix of linearly independent m nodes and n branches; x_τ – a vector of flow rates in the network segments t/h; G_τ –a vector of mass flow rates at nodes, t/h; $\overline{\mathrm{A}}^{\mathrm{T}}$ – a transposed coupling matrix of m nodes and n branches; $\overline{\mathrm{P}_\tau}$ – a vector of nodal pressures, mwc; h_τ – a vector of losses, mwc; H_τ – a vector of real heads, mwc; S and X_τ $(n\times n)$ – diagonal matrices of order n, including the values of coefficients of hydraulic resistance of branches s (mh^2/t^2) and absolute values of flow rates in them $|x_\tau|$ (t/h).

The equation relating the quantity of heat with mass flow rate of heat carrier is determined by the known relationship [10]:

$$Q_{j\tau} = c \cdot G_{j\tau} \cdot (t_1 - t_2), \quad (4)$$

where c – heating capacity of heat carrier, GJ/kg· °C; t_1 and t_2 – temperature of the network water in supply and return pipelines, respectively, °C;

Part of nodes in the hydraulic circuit represent the sites where heat sources that have heat output limitations are located, GJ/h:

$$Q^{hs}_{j\_\min} \le Q^{hs}_{j\tau} \le Q^{hs}_{j\_\max},\ j \in J_{hs}, \tau = \tau_0, ..., T. \quad (5)$$

A set of consumers in the heat market can be represented by two subsets: residential consumers and industrial consumers.

Demand of residential consumers for heat $Q^{rc}_{j\tau}$ is determined by the curve of heating load duration. Configuration of this curve is rather well defined by Rossander equation. According to this equation, heating load at each time τ is determined by the expression [11], GJ/h:

$$Q^{rc}_{j\tau} = \left[1 - (1-r) \cdot \left(\tau/\tau_{hp}\right)^{(g-r)/(1-g)}\right] \cdot Q^{hl}_j + Q^{hwl}_j,\ j \in J_{rc} \quad (6)$$

where Q^{hl}_j – calculated heating load, GJ/h; Q^{hwl}_j – calculated hot water load, GJ/h; r and g – coefficients of heating load curve unevenness; γ – share of hot water load; τ_{hp} – duration of heating period.

Demand for heat of the industrial consumer $Q^{IC}_{j\tau}$ is modeled by demand characteristic. This characteristic is based on real calculations for individually considered industrial consumers by approximating the retrospective data, given the forecast assessments of heat consumption volumes and heat prices. A more general form of the characteristic can be as follows [12, 13], GJ/h:

$$Q^{ic}_{j\tau} = \varphi_j(w_{j\tau}),\ j \in J_{ic}, \quad (7)$$

where $\varphi_j(w_{j\tau})$ – demand for heat of the consumer.

Each industrial consumer has their limitations on heat consumption, GJ/h:

$$Q^{ic}_{j\tau} \le Q^{ic}_{j\_\max},\ j \in J_{ic}. \quad (8)$$

Behavior of the unified heat supply organization in the heat market is determined by the amount of profit to be gained as a result of meeting the demand for heat specified by the consumer. In this case, the objective function takes account of both the heat production costs, and the costs related to its transportation to each consumer [14], EUR:

$$P^{UHSO} = \sum_{j \in J_{hs}} w_\tau \cdot Q^{hs}_{j\tau} - Z^{total}, \quad (9)$$

where $Z^{Total} = \sum_{j \in J_{hs}} Z^{hs}_j(Q^{hs}_j) + \sum_{i \in I} Z^{hn}_{i\tau}$ – total costs related to heat production $\sum_{j \in J_{hs}} Z^{hs}_j(Q_j)$ and heat transportation $\sum_{i \in I} Z^{hn}_{i\tau}$, EUR; w_τ – heat price or the inverse demand function, EUR/GJ.

The heat production costs of heat sources at any time τ can be represented by a functional dependence on the amount of heat produced by them. It includes both variable (fuel) and constant (maintenance) costs [15, 16]:

$$Z^{hs}_{j\tau}(Q^{hs}_{j\tau}) = \alpha_j \cdot (Q^{hs}_{j\tau})^2 + \beta_j \cdot Q^{hs}_{j\tau} + \gamma_j,\ \alpha_j > 0,\ j \in J_{hs}, \tau = \tau_0, ..., T, \quad (10)$$

where α_j, β_j, γ_j – approximation coefficients of heat source cost characteristic.

Heat network costs are determined according to the following analytical dependence [17] and include operating costs and costs of heat carrier pumping in the heat networks:

$$\sum_{i \in I} Z_{i\tau}^{hn}(x_{i\tau}) = F_1 + F_2 \cdot \sum_{i \in I} x_{i\tau}^2 |x_{i\tau}| \cdot s_i , \tag{11}$$

where $x_{i\tau}$ – heat carrier flow rate in the i–th section of the heating network at time τ, t/h; F_1 – semi-fixed costs, EUR; F_2 – coefficient at semi-fixed heat network costs, EUR.

Based on the above mentioned, an equilibrium between demand and supply in the heat market for liberalized model of the unified heat supply organization is determined by solving the problem of profit maximization of the organization [14], given constraints.

Find:

$$\sum_{\tau=\tau_0}^{T} P_\tau^{UHSO} = \sum_{\tau=\tau_0}^{T} \sum_{j \in J_{hs}} w_\tau \cdot Q_{j\tau}^{hs} - \sum_{\tau=\tau_0}^{T} \sum_{j \in J_{hs}} Z_{j\tau}^{hs}(Q_{j\tau}^{hs}) - \sum_{\tau=\tau_0}^{T} \sum_{i \in I} Z_{i\tau}^{hn}(x_{i\tau}) \to \max, \tag{12}$$

subject to (1)-(8).

The search for an optimal solution to the developed mathematical model of the unified heat supply organization is based on the univariate relaxation method [18] with the subsequent use of methods of redundant design [19] and simple iterative schemes. The method suggests the reduction of the multidimensional optimization problem to one-dimensional one and the use of a stepwise procedure for the improvement of solutions concerning the volumes of heat production by heat sources.

The computational process of search for an equilibrium between demand and supply based on the considered model can be presented graphically. To this end, we will consider a heating system with two heat sources and time interval $\tau = 1$ (Fig.1).

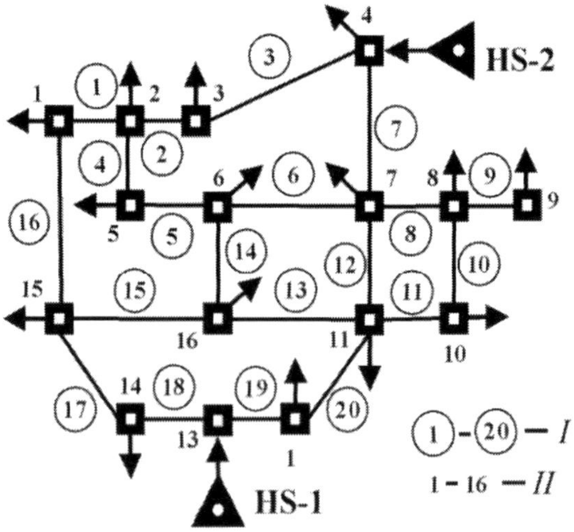

I – numbers of heat network sections; *II* – numbers of heat network nodes
Fig.1. A calculated scheme of heat supply to consumers.

Graphically, the objective function of the unified heat supply organization is a convex paraboloid with its vertex directed upward (Fig.2 a)). Figure 2 b) illustrates isolines of an objective function that show the area of change in the profit under various combinations of heat production volumes of heat sources. The computational process starts with a zero coordinate toward an increase in the profit of the unified heat supply organization (point A, Figure 2 b). Optimal solution for the considered example is obtained already in the 5th iteration and corresponds to point Q^*. The broken line A-B-C-D-Q^* represents the computational process trajectory itself.

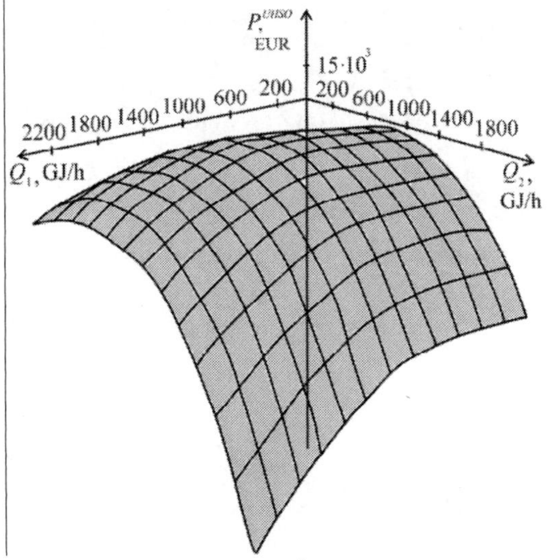

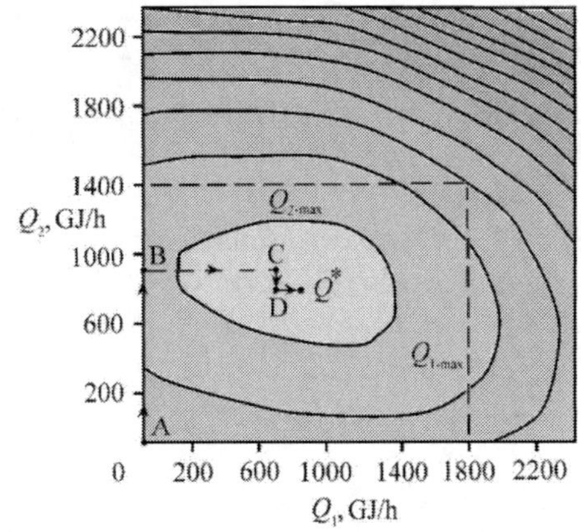

Fig.2. Graphical interpretation of a computational process of search for an optimal solution.

The equilibrium achieved at point Q^*, represents an optimal solution in terms of the heat volumes produced by the heat sources, equilibrium price of heat production and profit of the unified heat supply organization that corresponds to these values, provided the physical and technical constraints in the heating system are met, and demand equals supply at time $\tau = 1$ for the calculated ambient temperature.

3 Conclusions

An equilibrium mathematical model is developed for the organizational model of the heat market in the format of the unified heat supply organization. The model is based on the microeconomic model of the monopoly market. This mathematical model makes it possible to take into account the heat production and transportation costs as a single economic criterion. An algorithm is proposed to search for an optimal solution to the mathematical model of the unified heat supply organization. The algorithm is based on the method of univariate relaxation. The heating system with two heat sources was used to exemplify the computation technique of search for a solution.

4 Acknowledgements

The research was performed at Melentiev Energy Systems Institute SB RAS in the framework of a scientific projects III.17.4.3, № AAAA-A17-117030310437-4 of the foundation researches program of SB RAS.

References

1. L.D. Gitelman, V.E. Ratnikov, *Energy business* (Delo, Moscow, 2006) [in Russian]
2. V.V. Kudryavyi, Energ. Prom-st. Rossii, 2 (2015) [in Russian]
3. Federal Law of the Russian Federation «On heat supply»,190 (2010) [in Russian]

4. V.P. Busygin, E.V. Zhelobodko, A.A Tsyplakov, *Microeconomics – the third level* (Novosibirsk State University, Novosibirsk, 2003) [in Russian]
5. J. Tirole, *The Theory of Industrial Organization* (MIT Press., Cambridge, MA, 1988)
6. D. Carlton, J. Perloff. *Modern Industrial Organization.* (Reading, MA, Addison-Wesley, 2000)
7. P .Belleflamme, M. Peitz, *Industrial Organization: Markets and Strategies* (Cambridge University Press, Cambridge, UK, 2010)
8. A.P. Merenkov, V.Y. Khasilev, *Theory of hydraulic circuits* (Nauka, Moscow, 1985) [in Russian]
9. A.P. Merenkov, *Mathematical modeling and optimization of heat, water, oil and gas supply system* (Nauka, Novosibirsk, 1992) [in Russian]
10. E.Y. Sokolov. *District heating and heat networks* (Energoizdat, Moscow, 1982) [in Russian]
11. A. Rossander, *Die Anwendung der symbolischen Belastungskurven für die Elektrizitätswerke* (ETZ, 1913) [in German]
12. P. Vasant, N. Voropai, *Sustaining power resources through energy optimization and engineering* (Engineering Science Reference, Hershey PA, USA, 2016)
13. V.A. Stennikov, O.V. Khamisov, A.V. Penkovsky, International Journal of Energy Optimization and Engineering, 2 (2013)
14. H. Gravelle, R. Rees, *Microeconomics, 3rd Edition,* (Prentice Hall, Essex, England, 2004)
15. V.A. Stennikov, O.V. Khamisov, A.V. Penkovsky, Thermal Engineering, **58 (12)**, 1043, (2011)
16. A.V. Penkovskii, V.A. Stennikov, O.V. Khamisov, Thermal Engineering, **62 (8)**, 598, (2015)
17. A. V. Penkovskii,V.A. Stennikov, O.V. Khamisov, E.E. Mednikova, I.V. Postnikov, Energy procedia, 105 (2017)
18. E. Shoup, *A practical guide to computer methods for engineers* (Prentice-hall, England, 1979)
19. E.V. Sennova, V.G. Sidler, *Mathematical modeling and optimization of developing heating systems* (Nauka, Novosibirsk, 1987) [in Russian]

Modeling of a Prosumer in Intelligent Integrated Energy Systems

Evgeny Barakhtenko[1,*], *Dmitry* Sokolov[1], and *Veronika* Shelekhova[1]

[1]Melentiev Energy Systems Institute of Siberian Branch of the Russian Academy of Sciences (ESI SB RAS), Pipeline Energy Systems Department, 130, Lermontov str., Irkutsk, Russia, 664033

Abstract. The integration of many energy systems of various types (electricity, heat, cooling, etc.) into a single technological entity involves the use of new information and communication technologies and a holistic multi-level automated control system, which provides the establishment of integrated intelligent energy systems. In such systems, the consumer can both provide themselves with various types of energy, and supply energy into the energy system. The paper presents a methodology for modeling the prosumer in intelligent integrated energy systems. The methodology includes the following components: a mathematical model of the prosumer in intelligent electricity, heat/cooling systems; a method of choosing the energy sources to cover their load with minimal costs of electricity, heat/cooling generation. The methodology developed for modeling the prosumer in intelligent electric, heat/cooling systems was implemented in software. The presented results are tested on a test energy supply scheme.

1 Introduction

The integration of many separate systems into a single technological complex ensures the implementation of new functional capabilities, the use of more advanced technologies in operation, and the construction of integrated intelligent systems with coordinated control of their operation, as well as active participation of the consumer in the energy supply process [1-5]. In such energy systems, the end user is considered as a partner of energy entities in terms of ensuring reliable operation of the energy system, acquires the status of an active one and becomes one of the main elements in the energy system [6]. These consumers can independently set the requirements of the amount of energy to be received, its quality and consumer properties, and energy services [7-10].

Design of the load-controlled consumption systems, including heterogeneous energy sources and storage devices usually involves solving a mixed integer linear programming problem [11-12]. Some researchers [13-16] use multi-criteria optimization to consider not only the investment and operational costs in the design of energy systems, but also compensation for the energy generation-related emissions of pollutants.

The switch to the intelligent energy industry based on a customer-oriented approach generates the need to develop a methodological support for the study of prosumer

* Corresponding author: barakhtenko@isem.irk.ru

consumption, which is a pressing scientific problem of great importance for the national energy development. In this paper, we propose an approach to modeling of the prosumer. This approach makes it possible based on a redundant energy supply scheme to choose an optimal mix of generating equipment to meet the consumer needs for various types of energy, given its transport.

2 Statement of the problem

The set parameters include: a set of electricity sources $I = I_\mathrm{E} \cup I_\mathrm{N}$, consisting of sets of existing I_E and new I_N ones; a set of heat sources $J = J_\mathrm{E} \cup J_\mathrm{N}$, consisting of sets of existing J_E and new J_N ones ; a set of cooling energy sources, consisting of sets of existing D_E and new D_N ones , here $J \subset D$; a set of combined heat and power plants (CHPP) $CHP = (I_\mathrm{E}, J_\mathrm{E}),\ I_\mathrm{E} \subset I,\ J_\mathrm{E} \subset J$; the cost of primary energy resources to produce electricity $S^{\mathrm{EL}}_{t,i},\ i \in I$, heat $S^{\mathrm{HE}}_{t,j},\ j \in J$ and cooling energy $S^{\mathrm{CO}}_{t,d},\ d \in D$ at time *t*; demand for electricity C^t_{EL}, heat C^t_{HE} and cooling energy C^t_{CO} at time *t*; distance to the source of electricity $l^{\mathrm{EL}}_i,\ i \in I$, heat $l^{\mathrm{HE}}_j,\ j \in J$ and cooling energy $l^{\mathrm{CO}}_d,\ d \in D$. Solution to the problem should include: electricity generation $\mathbf{w} = (w_1,\dots, w_i)^\mathrm{T}$; heat generation $\mathbf{h} = (h_1,\dots, h_j)^\mathrm{T}$; cooling generation $\mathbf{q} = (q_1,\dots, q_d)^\mathrm{T}$. It is necessary to minimize the function of total costs of the load-controlled consumer energy system that has the form:

$$Z(w,\ h,\ q) = \sum_{t=1}^{p} \left\{ \begin{array}{l} \sum_{m=1}^{i} \alpha_m \Big[Z_t^{\mathrm{EL}}\left(w_m, S^{\mathrm{EL}}_{t,m}\right) + Z^{\mathrm{S\_B}}\left(w_m\right) + Z_t^{\mathrm{S\_E}}\left(w_m\right) + \\ + Z^{\mathrm{N\_B}}\left(w_m, l^{\mathrm{EL}}_m\right) + Z_t^{\mathrm{N\_E}}\left(w_m, l^{\mathrm{EL}}_m\right)\Big] + \sum_{n=1}^{j} \beta_n \Big[Z_t^{\mathrm{HE}}\left(h_n, S^{\mathrm{HE}}_{t,n}\right) + \\ + Z^{\mathrm{S\_B}}\left(h_n\right) + Z_t^{\mathrm{S\_E}}\left(h_n\right) + Z^{\mathrm{N\_B}}\left(h_n, l^{\mathrm{HE}}_n\right) + Z_t^{\mathrm{N\_E}}\left(h_n, l^{\mathrm{HE}}_n\right)\Big] + \\ + \sum_{r=1}^{d} \gamma_r \Big[Z_t^{\mathrm{CO}}\left(q_r, S^{\mathrm{CO}}_{t,r}\right) + Z^{\mathrm{S\_B}}\left(q_r\right) + Z_t^{\mathrm{S\_E}}\left(q_r\right) + \\ + Z^{\mathrm{N\_B}}\left(q_r, l^{\mathrm{CO}}_r\right) + Z_t^{\mathrm{N\_E}}\left(q_r, l^{\mathrm{CO}}_r\right)\Big] \end{array} \right\} \to \min \qquad (1)$$

$$\boldsymbol{\alpha} = \left(\alpha_1,\dots,\alpha_i\right)^\mathrm{T},\ \alpha_m \in \{0,1\},\ \boldsymbol{\beta} = \left(\beta_1,\dots,\beta_j\right)^\mathrm{T},\ \beta_n \in \{0,1\},\ \boldsymbol{\gamma} = \left(\gamma_1,\dots,\gamma_l\right)^\mathrm{T},\ \gamma_r \in \{0,1\},$$

$$\mathbf{w} = \left(w_1,\dots,w_i\right)^\mathrm{T},\ w \in \mathbb{R}^i,\ \mathbf{h} = \left(h_1,\dots,h_j\right)^\mathrm{T},\ h \in \mathbb{R}^j,\ \mathbf{q} = \left(q_1,\dots,q_d\right)^\mathrm{T},\ q \in \mathbb{R}^d,\ S^{\mathrm{EL}}_{t,m} \in \mathbb{R}^{m\times p},$$

$$S^{\mathrm{HE}}_{t,n} \in \mathbb{R}^{n\times p}, S^{\mathrm{CO}}_{t,r} \in \mathbb{R}^{r\times p},\ l^{\mathrm{EL}}_m \in \mathbb{R}^i,\ l^{\mathrm{HE}}_n \in \mathbb{R}^j, l^{\mathrm{CO}}_r \in \mathbb{R}^d, t = \overline{1,p},\ m = \overline{1,i},\ n = \overline{1,j},\ r = \overline{1,d}.$$

Subject to:

$$w_i = w_{i0} K_i^{\mathrm{TR}} K_i^{\mathrm{EF}} \lambda_i,\ i \in I;\ h_j = h_{j0} K_j^{\mathrm{TR}} K_j^{\mathrm{EF}} \lambda_j,\ j \in J;\ q_d = q_{d0} K_d^{\mathrm{TR}} K_d^{\mathrm{EF}} \lambda_d,\ d \in D; \quad (2)$$

$$C^t_{\mathrm{EL}} \le \sum_{m=1}^{i} w^t_m;\ C^t_{\mathrm{HE}} \le \sum_{n=1}^{j} h^t_n;\ C^t_{\mathrm{CO}} \le \sum_{r=1}^{d} q^t_r; \quad (3)$$

$$w \ge 0;\ h \ge 0;\ q \ge 0; \quad (4)$$

$$\underline{w} \le w_i \le \overline{w},\ i \in I;\ \underline{h} \le h_j \le \overline{h},\ j \in J;\ \underline{q} \le q_d \le \overline{q},\ d \in D; \quad (5)$$

$$\underline{\eta} \le \frac{w_v}{h_v} \le \overline{\eta},\ v \in I_\mathrm{E}, I_\mathrm{E} \subset I, v \in J_\mathrm{E},\ J_\mathrm{E} \subset J; \quad (6)$$

where $Z_t^{EL}\left(w_m, S_{t,m}^{EL}\right)$, $Z_t^{HE}\left(h_n, S_{t,n}^{HE}\right)$, . $Z_t^{CO}\left(q_r, S_{t,r}^{CO}\right)$. – calculated costs of electricity, heat and cooling, respectively; $Z^{S\_B}\left(w_m\right)$, $Z^{S\_B}\left(h_n\right)$, $Z^{S\_B}\left(q_r\right)$ – calculated construction costs of the source of electricity, heat and cooling, respectively; $Z_t^{S\_E}\left(w_m\right)$, $Z_t^{S\_E}\left(h_n\right)$, $Z_t^{S\_E}\left(q_r\right)$ – calculated operational costs of the source of electricity, heat and cooling, respectively; $Z^{N\_B}\left(w_m, d_m^{EL}\right)$, $Z^{N\_B}\left(h_n, d_n^{HE}\right)$, $Z^{N\_B}\left(q_r, d_r^{CO}\right)$ – calculated construction costs of a network from the source of electricity, heat and cooling , respectively, to the consumer; $Z_t^{N\_E}\left(w_m, d_m^{EL}\right)$, $Z_t^{N\_E}\left(h_n, d_n^{HE}\right)$, $Z_t^{N\_E}\left(q_r, d_r^{CO}\right)$ – calculated operational costs of the network from the source of electricity, heat and cooling , respectively, to the consumer; p – period of time; α_m – electricity source state (0 – off, 1 – on); β_n – heat source state (0 – off, 1 – on); γ_r – cooling source state (0 – off, 1 – on); K^{TR} – the conversion factor of the primary energy resource to the required energy type; K^{EF} – source efficiency; λ – energy transmission losses in the energy system components; w_{i0} – primary energy resources; η – generated electricity to heat ratio at CHPP.

The above generalized statement was reduced to solving the problem of mixed integer linear programming [17-18].

3 Testing on a test energy scheme

A redundant energy supply scheme for the prosumer is built to substantiate the choice of energy sources of the prosumer to cover its own load with minimal costs for generation of electricity, heat and cooling. The redundant scheme consists of a coal-fired mini-CHPP (CCHPP), a gas-fired mini-CHP (GCHPP), an electric boiler, a heat storage, and an absorption refrigeration unit. Generating capacities include an Ingersoll Rand microturbine, an absorption refrigeration unit (ARU), a Neuron-500 container power plant unit, and an electric boiler. The generation structure of the active consumer is shown in Figure 1.

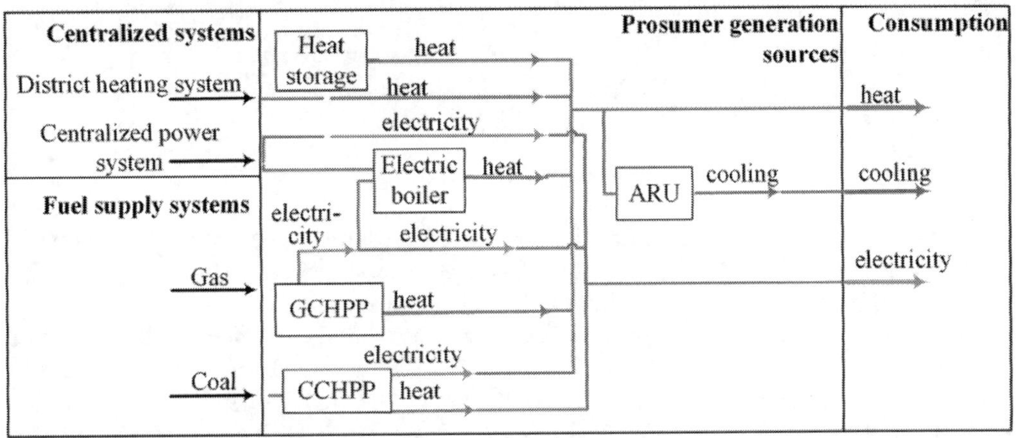

Fig. 1. A structure of controlled load consumer generation.

According to the above scheme, the sources of energy supply to the prosumer in the considered scheme are the following: the electricity sources are: centralized power system (CPS), CCHPP, GCHPP; heat sources are: district heating system (DHS), electric boiler, CCHPP, heat storage (HS), GCHPP; cooling sources are: ARU that receives heat from heat sources.

The constructed daily electricity, heat and cooling load curves of the prosumer are presented in Figure 2.

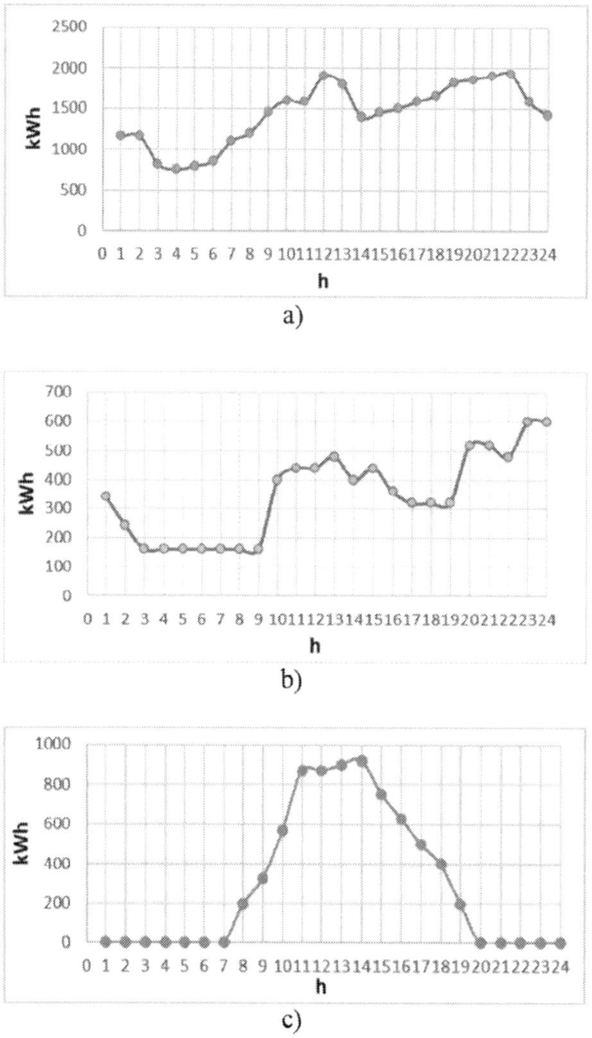

Fig. 2. Daily load curve: a) electricity, b) heat and c) cooling.

The costs of various resources in the considered prosumer scheme are the following: CPS – 0.92 rub./ kWh, DHS – 0.89 rub./ kWh, gas - 8.49 rub./m^3, coal – 0.11 rub./kg, storage in HS – 0.01 rub./m^3. The model takes into account the following energy transmission and storage losses in the energy system: 5% when stored in HS, 7% when transmitted to DHS, 7% when transmitted to CPS. The assumed constraints include the constraints on electric boiler capacity: $h \leq 1500$ kW; on the GCHPP capacity: 12.5 kW $\leq w \leq$ 2000 kW, 18 kW $\leq h \leq$ 4000 kW, $1.44 \leq w / h \leq 2$; on the CCHPP capacity: 100 kW $\leq w \leq$ 1500 kW, 192 kW $\leq h \leq$ 2900 kW, $1.92 \leq w / h \leq 1.93$; on HS capacity: 30 kW $\leq h \leq$ 583 kW.

We have developed the software to model the prosumer. The software is implemented in Java [19]. A set of tools implemented as a free lp_solve library for Java [20] is employed as a solver of linear programming problems.

The results of calculation for the selection of the prosumer generation sources in an intelligent electricity, heat/cooling system are presented in Figure 3. The following technologies were selected from the competing ones: CPS, GCHPP, CCHPP – for electricity generation; an electric boiler, GCHPP, CCHPP – for heat generation; and the use of heat from GCHPP and CCHPP – for cooling generation. Two sources are selected for cooling generation. They supply heat to ARM. The CCHPP is the most cost-effective source, but due to the limitations on the minimum possible heat supply at some hours of the day (8-00 and 19-00, see Figure 3), this source cannot be used. In this case, the GCHPP is used to cover the cooling load. The total costs of electricity, heat and cooling generation were 114864.15 rubles.

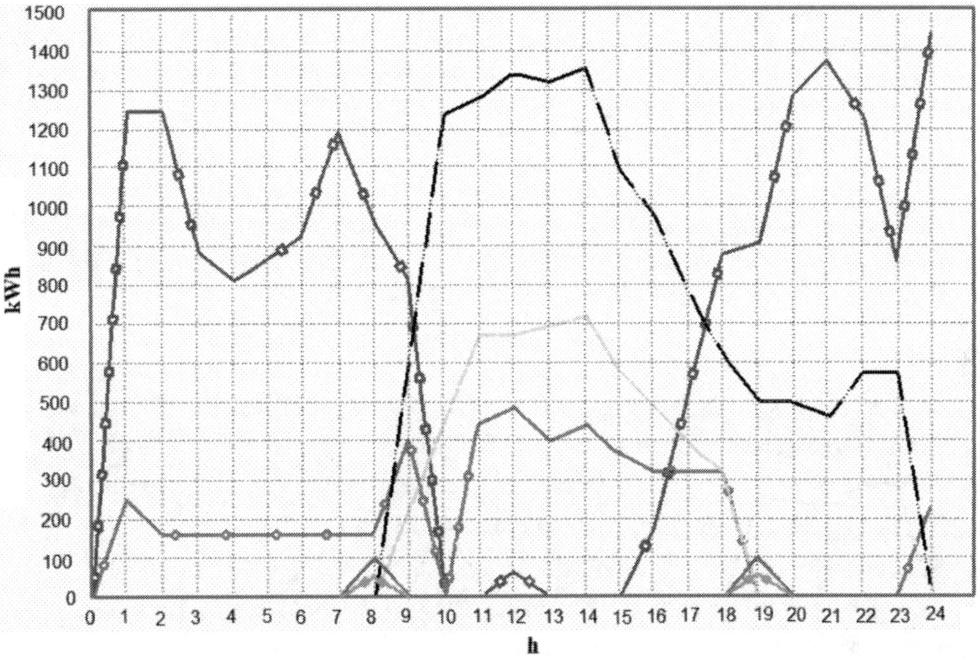

Fig. 3. The result for the determination of the rational loading of the prosumer generation sources (electric boiler, GCHPP (heat), CCHPP (heat), DHS, CPS, GCHPP (electricity), CCHPP (electricity), GCHPP (cooling), CCHPP (cooling)).

4 Conclusion

In this research, we developed a methodology for modeling the prosumer in intelligent electricity and heat supply systems.

The most important results of the work are: a mathematical model of the prosumer in the intelligent electricity, heat and cooling systems; a method for rational management of electricity, heat and cooling generation of the prosumer in the intelligent systems to cover the load with minimal costs; a method of choosing the energy sources to cover the load with minimal costs for electricity, heat and cooling generation; software implementation of the developed methodology for the modeling of the prosumer in intelligent electric, heat/cooling systems.

The further development and extension of the started research will focus on: consideration of the wear and tear of the generating equipment during its start-up/shutdown; consideration of nonlinear characteristics of equipment; consideration of the inertia nature of the processes that occur in the heat and cooling systems.

The work was carried out within framework of scientific projects III.17.4.1 of the fundamental research program of the SB RAS AAAA-A17-117030310432-9.

References

1. N. I. Voropai and V. A. Stennikov, Izvestiya Rossiiskoi akademii nauk. Energetika, 1 (2014)
2. N. Voropai, V. Stennikov, S. Senderov, E. Barakhtenko, O. Voitov, A. Ustinov, Journal of Energy Engineering, **143** (2017)
3. N. I. Voropai, V. A. Stennikov, S. M. Senderov, E. A. Barakhtenko, L. I. Kovernikova, O. N. Voitov, T. B. Oshchepkova, L. V. Semenova, Energeticheskaya Politika, 3 (2015)
4. N. Voropai, V. Stennikov, S. Senderov, E. Barakhtenko, O. Voitov, L. Kovernikova, T. Oshchepkova, L. Semenova, *Proc. of the International Conference on Problems of Critical Infrastructures*, 115-121 (2015)
5. N. I. Voropai, V. A. Stennikov, E. A. Barakhtenko, Studies on Russian Economic Development, **28**, 492–499 (2017)
6. H. Yang, T. Xiong, J. Qiu, D. Qiu, Z. Y. Dong, Applied Energy, **167**, 353-365 (2016)
7. R. Zafara, A. Mahmoodb, S. Razzaqc, W. Alia, U. Naeema, Renewable and Sustainable Energy Reviews, **82**, 1675-1684 (2018)
8. D. Patteeuw, K. Bruninx, A. Arteconi, E. Delarue, W. D'haeseleer, L. Helsen. Applied Energy, **151**, 306-319 (2015)
9. N. Neyestani, M. Yazdani-Damavandi, M. Shafie-khah, G. Chicco, J. P. S. Catalão, IEEE Transactions on Smart Grid, **6**, 1748-1762 (2015)
10. S. van der Stelt, T. AlSkaif , W. van Sark, Applied Energy, **209**, 266-276 (2018)
11. E. Barbour, M. C. González, Applied Energy, **215**, 356–370 (2018)
12. S. Samsatli, N. J. Samsatli, Applied Energy, **220**, 893-920 (2018)
13. M. La Scala, A. Vaccaro, A. F. Zobaa. Applied Thermal Engineering **71**, 658-666 (2014)
14. R. Rezaeipour, A. Zahedi, Solar Energy, **157**, 227–235 (2017)
15. F. Samira, M. Pierre, B. Gwenaelle, M. Francois, Energy, **45**, 12-22 (2012)
16. J. Aghaei, Mohammad-Iman Alizadeh. Energy, **55**, 1044-1054 (2013)
17. M. J. Todd, Mathematical Programming, **91**, 417-436 (2002)
18. G. B. Dantzig, M. N. Thapa, *Linear Programming 2: Theory and Extensions* (2003)
19. J. Gosling, B. Joy, G. Steele, G. Bracha, A. Buckley, *The Java Language Specification* (2014)
20. lp_solve reference guide. URL: http://lpsolve.sourceforge.net/5.5/

Placement of measurements for electrical distribution network state estimation

Irina Golub[1*], and *Yana* Kuzkina[2]

[1]Melentiev Energy Systems Institute of Siberian Branch of Russian Academy of Sciences, Pipeline Energy Systems Department, Lermontov str., 130, Irkutsk, Russia, 664033
[2] LLC "Engineering Center "Irkutskenergo", Irkutsk, Russia

Abstract. The paper is concerned with the problem of placement of the minimum number of smart meters to ensure either the observability of all state variables in distribution network or the observability of voltage magnitudes. Voltage control is important in the distribution network with distributed generation sources which adoption can lead to unpredictable overvoltage exceeding admissible values. The algorithm for smart meters including measurements of voltage magnitudes and active and reactive current injections is similar to the algorithm of choosing the minimum number of phasor measurement units to ensure topological observability. Optimal control of the active distribution network operation requires monitoring to be based on a classical linear state estimation procedure. The results of the research demonstrate the effectiveness of the proposed approaches and are illustrated by example of a test distribution network.

1 Introduction

The enhancement of the energy efficiency can be achieved through the creation of intelligent networks including electric and pipeline distribution networks [1]. The main feature of active intelligent networks is the integration of a distribution network with a communication network and a state-of-the-art measurement system. Such integration became possible due to the development of new information and communication technologies and the creation of Advanced Metering Infrastructure (AMI) [2]. The most important AMI components are smart meters (SM) installed at load nodes. Their data are transferred to a utility or a dispatcher, thus providing a two-way data exchange between a meter, a central system and a consumer. Consumers knowing their energy consumption and prices can manage their energy demand which in turn can reduce the total energy consumption.

Intelligent systems have been widely implemented not only in the electric power industry, but also there is AMI for collection and transmission of information about water, gas, and heat consumption. In other words, electro-SM, water-SM, heat-SM, gas-SM [3, 4] apply the same devices for information transfer and the same approaches to storage and processing results of measurements. Moreover, the group of EU Commission experts considers it is necessary to create integrated intelligent networks in which electrical and pipeline networks will be linked together [4]. EU is planning to replace 80% of all types of

* Corresponding author: golub@isem.irk.ru

electric measurements in distribution networks by SM measurements by 2020 [5], the leaders in the installation of heat-SMs are Denmark, Sweden and China, and gas-SM – the UK and Italy [4].

Various communication tools can be used to create AMI, but the power line communication (PLC) technology found the widest application for electrical distribution networks. It provides high-speed data transfer throughout the power networks. Apart from the information about electricity consumption, SMs of some manufacturers can measure voltage magnitude, active and reactive power or current injections. As an example of such a SM we can consider a three-phase electronic device A1500. The device has a time marker with a measurement error not more than 0.5% and a 15-minute interval between measurements [6]. The voltage information is important for power quality control at certain nodes, meanwhile full synchronized information about current loads and voltages is used for state estimation of the distribution network.

However, currently there is no standard for transfer of SM measurements, and the studies on the problem of applying SMs in the distribution network state estimation do not provide an unambiguous answer to the question about the possibility of their synchronization. [2, 6, 7] Thus, the studies on the development of technologies that provide both phasor measurements in distribution network and their synchronization are conducted, and it is necessary to be ready to use them. In the analysis of algorithms for placing SMs in the distribution network, which are the focus of the paper, firstly SMs are supposed to measure phasors of current injections and voltage phasors, and then it is shown how the replacement of voltage phasor measurements with the measurements of voltage magnitudes affects the placement of SMs.

General approaches of the AMI organization for electrical and pipeline distribution networks allow to use similar algorithms for the placement of SMs. In the future, indications of such SMs will be used not only for energy consumption control, but also for the state estimation of the relevant pipeline networks

2 Mathematical statement

Optimal control of the active distribution network operation requires monitoring to be based on a classical linear state estimation procedure involving synchronized PMU (phasor measurement unit) measurements [9].

State estimation on the basis of PMU measurements suggests determining components of the state vector x, which, in an algebraic form of complex values, includes real U_a and imaginary U_r voltage components at the nodes of the calculated electrical network scheme on the basis of synchronized phasor measurements z, including measurements of voltage phasors $U\angle\delta_U = U_a + jU_r$ at nodes with PMU and phasors of currents $I\angle\delta_I = I_a + jI_r$ in the ties adjacent to such nodes. Measurements z are related to the estimated variables of the state vector x by the Jacobian matrix H. The elements of this matrix are parameters of the electrical network equivalent circuit; such a relation is linear

$$z = H \cdot x \quad (1)$$

Equality of the rank of a matrix H to the number of state variables x is a necessary condition for observability, which determines the existence of a solution to problem (1). Since the rank of the matrix H largely depends on the network topology, the set and allocation of measurements, such observability is called topological.

With the PMU measurements used, node i will be topologically observable, if voltage phasor is measured at this node, or voltage phasor at node i can be determined by the measurements of voltage phasor at the adjacent node j and current phasor in the tie line i-j.

Additional information used to provide the observability is the information on equality of current injections to zero at transit nodes, i.e. at the nodes without generation and load. If at the transit node i voltage phasor is known, then zero injection at node i can be used to provide observability of any node j adjacent to node i. The zero-injection node i will be observable if voltage at any node j adjacent to it is known.

In the distribution network the synchronized PMU measurements are not applied due to their high price [3]. Nevertheless, to place SMs in the distribution network we can apply the same approaches as to place the minimum number of PMUs [10] with the use of an integer linear programming procedure:

$$\min_{x} f^T, \qquad Ax \geq b \tag{2}$$

where f - vector of linear objective function coefficients, b - vector of right-hand sides for inequality constraints, A - a matrix, x - a binary integer vector which elements take values 0 or 1, the latter indicate which nodes should be set to PMU. A detailed study of the algorithm for choosing the optimal composition of measurements, including when limiting the number of channels for current measurements and zero current injections is given in [9-11]. The problem of choosing the minimum number of PMUs is solved for the normal operation condition of the network (Task 1), for the condition of change in the network topology when some individual ties are disconnected, (Task 2), when some PMUs fail (Task 3), and for the case where some of the voltage or current phasor measurements that belong to the PMU are lost (Task 4).

3 An example of a test distribution network

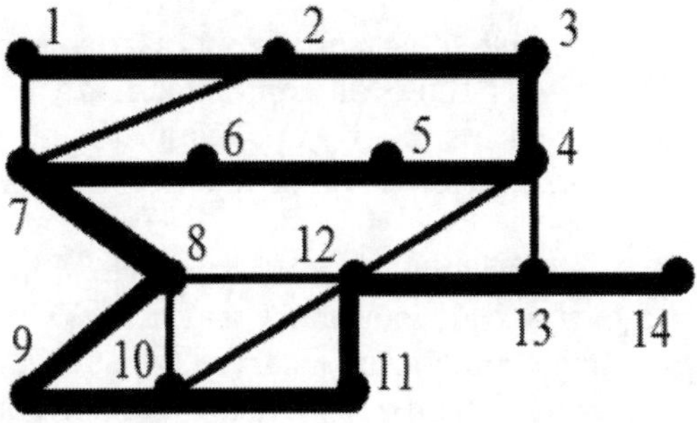

Fig.1. A 14-node IEEE test scheme with 20 ties, with the Hamiltonian path highlighted with a bold line

For the 14-node IEEE test scheme with 20 ties, Fig.1, we can show that the placement of one- or two-channel PMUs at it for Tasks 1-4 can be obtained when placing the PMUs at the nodes of the Hamiltonian path [12]. This way is demonstrated in Fig.1 by a bold line, it passes through nodes 1-14 entering each of them only once.

The placement one- and two-channel PMUs at the nodes of Hamiltonian path for Tasks 1 and 3 is demonstrated in Fig.2 and Fig.3. The number of one-channel PMUs for these Tasks equals 7 and 14 and the number of two-channel PMUs – 5 and 10, respectively.

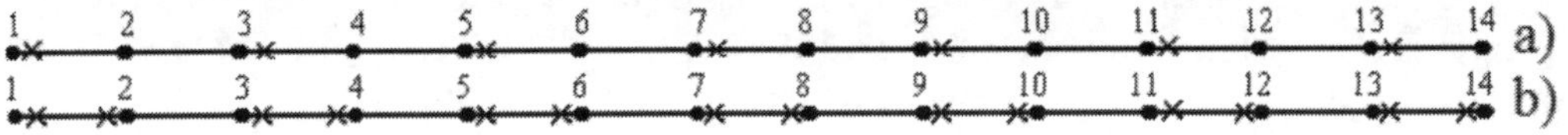

Fig.2. Sets of one-channel PMUs ⟶ for the network in Fig.1, including phasor measurements of current ⟶ and voltage ⟶ for Task 1 (a) and Task 3 (b)

The Hamiltonian path has a configuration typical for a low-voltage feeder without branching. Therefore, the solutions presented in Fig.2 and Fig.3 can be considered as optimal sets of PMUs providing solutions to Tasks 1 and 3 for the distribution network feeder.

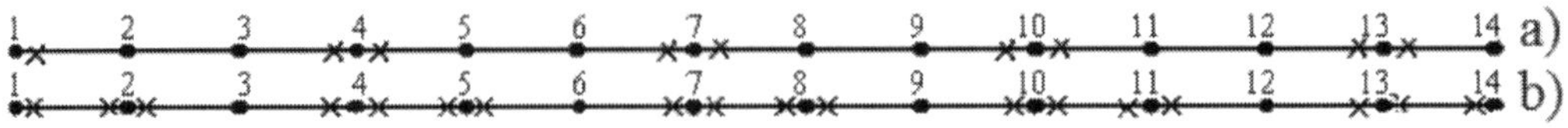

Fig.3. Sets of two-channel PMUs, for the network in Fig.2, including phasor measurements of current and voltage for Task 1 (a) and Task 3 (b)

4 An optimal set of SMs in the distribution network on the basis of PMU placement results

An analysis of possible selection of an optimal SM set by replacement of current measurements in feeder ties, Fig.2 and Fig.3, by the measurements of current injections shows that for Task 1 the optimal set of SM measurements can be obtained on the basis of placement of one-channel PMU, Fig.2a, for Task 3 – on the basis of placement of a two-channel PMU, Fig.3b. Such optimal sets of SMs are demonstrated in Fig.4.

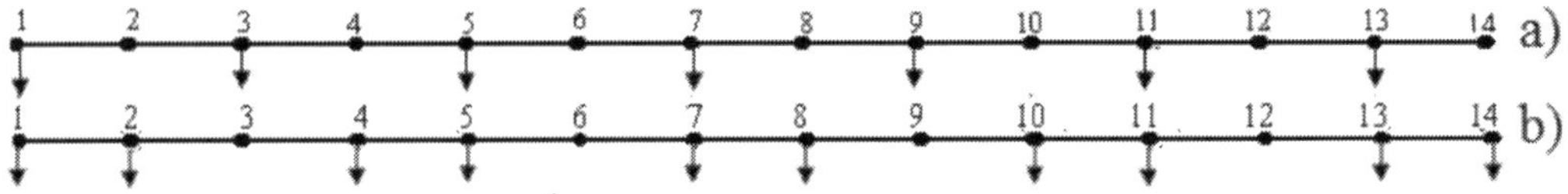

Fig.4. Sets of SM measurements for a 14-node distribution network without branches, including phasor measurements of current injections and voltage for Task 1 (a) and Task 3 (b)

As previously noted, it is possible to synchronize measurements of active and reactive currents and voltage magnitudes for the SMs applied currently. We will show how the replacement of phasor measurements of voltage with the measurements of voltage magnitudes affects the selection of a set of SMs that provides the distribution network observability.

The 0.4 kV low-voltage distribution network connected directly to consumers is characterized by a comparatively small inductance and a load with a high power factor. Therefore, the imaginary voltage component in the distribution network is considerably smaller than the real component. Experiments show that voltage phases do not exceed two degrees. This allows the state estimation of distribution network specify the measurements of real voltage components to be equal to the measurements of voltage magnitudes, and assume the measurements of imaginary voltage components to be equal to zero.

In the case where SMs with measurements of voltage magnitudes are installed at all load nodes of N-node network, the minimum number of measurements required to ensure observability should include $2(N-1)$ measurements of active and reactive current injections, a measurement of voltage magnitude at one of the nodes and a fixed voltage phase at the base node. The remaining $(N-2)$ measurements of voltage magnitudes will be redundant and can be used to increase the accuracy of the obtained estimates.

The diagrams presented in Fig.5 make it possible to compare the values of real voltage component for testing conditions of the 14-node low voltage feeder with the real voltage components that were obtained from linear state estimation for three sets of SMs.

The values of current injections and voltage magnitudes that were assumed to be equal to the measurements of real voltage components were obtained for the testing conditions and used as measurements for state estimation. The measurements of imaginary voltage components were set to be equal to zero.

The first set included SM measurements at nodes 2-14, the second one corresponds to Fig.4b (Task 3) and the third set corresponds to Fig.4a (Task 1). An analysis of the diagrams in Fig.5 shows that for all sets of SM measurements the real voltage components can be applied to control voltage levels at all the nodes of low-voltage feeder that is one of the main tasks in the distribution network with distributed generation.

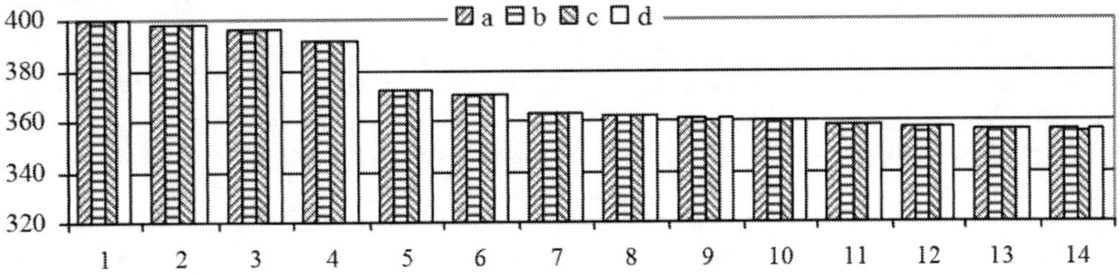

Fig.5. Values of real voltage components at nodes of a 14-node feeder: a – for the testing conditions, b – when SMs are installed at nodes 2-14, c – when SMs are installed according to Fig.4b, d – when SMs are installed according to Fig.4a.

The error in equating zero "measurements" of the imaginary voltage components leads to significant errors in the estimation of not only reactive but also active components of the currents in the branches and active current injections, as follows from the diagrams in Fig.6 and Fig.7. Thus because of errors in the estimates of currents the observability of all state variables can be provided only when SMs are installed at all load nodes.

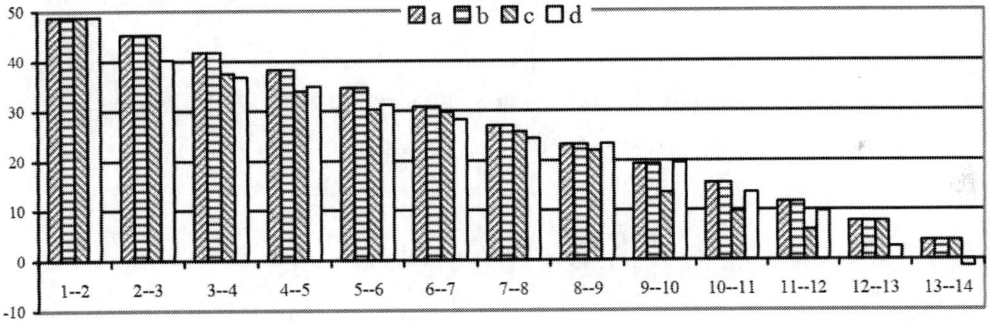

Fig.6. The values of the active components of the currents in branches of the 14th node feeder: a - for the test mode, b - when installing SM in nodes 2-14, c - when installing SM in accordance with Fig.4b, d - when installing SM in accordance with Fig.4a.

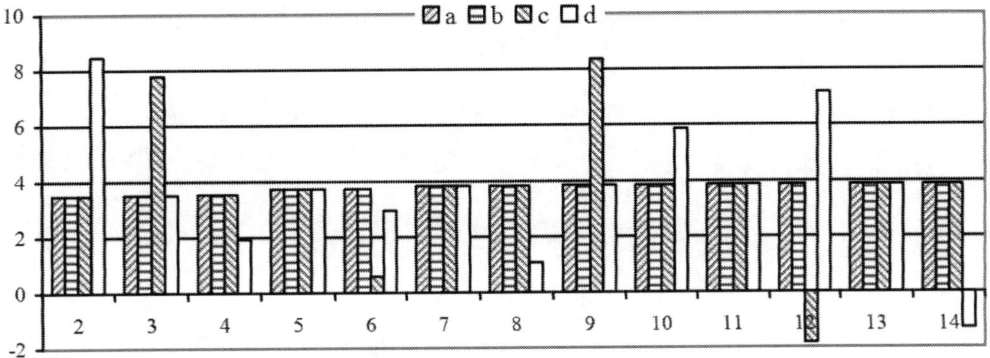

Fig.7. The values of the active current injections of the 14th node feeder: a - for the test mode, b -

when installing SM in nodes 2-14, c - when installing SM in accordance with Fig. 5b, d - when installing SM in accordance with Fig. 5a.

5 Conclusions

1. It is proposed to use the algorithms for selecting the minimum number of PMUs in different operating conditions for the choice of the number and placement of SM measurements in the distribution network.

2. The minimum number of SM measurements for the normal operation condition of the network and when some SM measurements fail can be obtained by selection of one-channel and two-channel PMUs.

3. Replacement of phasor measurements of voltage in the sets of SM measurements with the measurements of voltage magnitudes makes it possible to provide observability real voltage components at all nodes of the distribution network.

4. Observability of all variables in a low-voltage distribution network can be provided only when SMs are installed at all load nodes.

5. Approaches similar to those proposed can be used to form SM formulations as separate pipeline distribution networks and to combine them with electric distribution networks.

The work is done in the framework of the project III.17.4.2. program of fundamental research SB RAS, registration number AAAA-A17-117030310438-1.

References

1. R.Wennersten, Q. Sun, H. Li, J CLEAN PROD, **103**, 13 (2015)
2. R.R. Mohassel, A. Fung, F. Mohammadi, K. Raahemifar, CCECE, 8 (2014)
3. S. Santos, A Llano, A. Arzuaga, T.Arzuaga, L.Marron, M.Zamalloa, CIGRE, 8 (2012)
4. Q. Sun, H Li., Z. Ma, C. Wang, J. Campillo, Q. Zhang, F. Wallin, J. Guo. IEEE Io Journal. **3**, 16 (2016)
5. Q. Chen, D. Kaleshi, Zh. Fan. IEEE ISGT EUROPE. 5 (2013)
6. W.H. Wellssow, D.Waeresch, PSCC. 7 (2014)
7. S. Rohrenbeck, IEEE PESS. 6 (2015)
8. Abart, D. Burnier, A. Lugmaier, CIRED. 4 (2011)
9. R.F. Nuqui, A.G Phadke, IEEE Power Tech. 4 (2007)
10. I.I. Golub, M.V. Khokhlov. Elektrichestvo **1**, 8 (2015)
11. N. Fan, J-P Watson, Energy Systems, **1**, 19 (2014)
12. Dharwadker, *The Hamiltonian circuit algorithm* (Amazon, 2011)

PLANNING OF THERMAL-HYDRAULIC TESTS OF PIPELINE SYSTEMS WITH THE METHODS OF ACTIVE IDENTIFICATION*

Oksana A. Grebneva[1,†]

[1] Melentiev Energy Systems Institute of Siberian Branch of the Russian Academy of Sciences (ESI SB RAS), Pipeline Energy Systems Department, 130, Lermontov str., Irkutsk, Russia, 664033

Abstract. The lack of reliable information about actual characteristics and parameters is the main deterrent to the effective application of mathematical and computer-modelling methods for solving problems of optimal reconstruction, adjustment, development of operational condition, dispatching control of pipeline systems. The paper presents the original formalization of the problems of testing pipeline systems for hydraulic and thermal losses as problems of active identification, assuming optimal planning and processing of experimental results on the basis of attracting a mathematical model of steady-state thermal-hydraulic state. The proposed technique consists in a stepwise testing strategy that ensures the extraction of maximum information with a minimum risk of excessive testing, is applicable for different types of testing of pipeline systems of arbitrary structure and configuration. Theoretically, and for an illustrative example, the potential efficiency of combining thermal and hydraulic tests of pipeline systems is shown, which is manifested in minimizing the total number of experiments to obtain a given accuracy in determining the actual characteristics of pipeline systems and the predictive properties of the model.

1 An analysis of literature and problem statement

The industrial methods to carry out both hydraulic and thermal tests [1-3] have some flaws: 1) by virtue of their laboriousness, the tests are normally done only for characteristic sections in terms of dates and operating conditions, which however does not guarantee a complete picture of the network state as a whole; 2) limitation on the time of the tests is related to some preparatory work to be performed before the tests (disconnection of some branches, installation of jumpers, etc.), for example, for heat network the tests are carried out during non-heating period; 3) conditions for the tests are poorly regulated; 4) the problem of placement of measuring devices, their number and mix is not formalized; 5)

* The research was carried out within the project III.17.4.3 of the Fundamental research program of SB RAS (AAAA-A17-117030310437-4) with finance support of RFBR and the Government of Irkutsk Region in the framework of research project № 17-48-380021

† Corresponding author: oksana@isem.irk.ru

hydraulic and thermal characteristics are determined not for each section individually but for the groups of sections (the obtained data are extrapolated to all sections); 6)The methods do not suggest determining the resistance of the consumer that are necessary for adequate modeling of operating conditions; 7) moreover, hydraulic and thermal tests are conducted separately, which is associated with additional material, labour and time expenditure.

In order to avoid all of these problems, we propose planning of joint optimal thermal-hydraulic tests that are aimed at simultaneously obtaining both hydraulic and thermophysical characteristics.

The joint thermal-hydraulic tests can potentially reduce laboriousness (decrease in the number of involved operating conditions, reduction in the number of measurements to be made) and time of the tests, as well as improve the accuracy of the parameters obtained as a result of the tests.

In many studies, such problems are solved by the methods of passive identification and mainly for the water supply systems [4-9].

The methodological framework for these studies is represented by the theory of hydraulic circuits developed at Melentiev Energy Systems Institute SB RAS [10]. The theory is considered as an intersectoral scientific and engineering discipline providing a single vector-matrix language and models to define a wide range of pipeline systems of various types and purposes. This theory was employed to formulate the task of active identification in pipeline systems and develop a new methodology [11-14]. Previously, the effectiveness of this methodology was studied when applied to the planning of hydraulic tests. The paper presents the studies on its application to plan joint thermal-hydraulic tests with an extended model of nonisothermal flow distribution.

2 Object, goal and objectives of the research

The primary goal of the research is to adapt the algorithms and methods for active identification to plan joint thermal-hydraulic tests of pipeline systems. The following objectives were set: 1) to modify the previously developed active identification planning method, in order to enable its application in planning of joint thermal-hydraulic tests of pipeline systems; 2) to demonstrate: a) the effectiveness of the active identification methods; b) the effectiveness of joint thermal and hydraulic tests – on a case study.

Mathematical statement of the problem. The planning of the tests is mainly aimed at influencing the major factors that affect the accuracy of the obtained parameters and characteristics of pipeline systems (hydraulic resistance coefficients and heat transfer coefficients). Moreover, for the tests to be feasible, it is necessary to take into account the constraints imposed on: 1) the state variables (including those non-fixed) that should not go beyond the technologically feasible limits after control actions; and 2) the amounts of measuring devices and sites for their possible installation. The problem is solved in the space of varied parameters.

In a general form the problem of experiment planning can be represented as follows [11,12]:

$$\min_{N,\boldsymbol{\delta},\mathbf{X}_{\mathbf{R}}} \{\det[\mathbf{C}_{\alpha}(\mathbf{X}_{\mathbf{R}}^{(1)},\dots,\mathbf{X}_{\mathbf{R}}^{(N)},\boldsymbol{\delta}^{(1)},\dots,\boldsymbol{\delta}^{(N)},\boldsymbol{\alpha})]\} \tag{1}$$

subject to

$$\underline{\mathbf{R}}^{(u)} \le \mathbf{R}^{(u)} = \begin{pmatrix} \mathbf{Y}^{(u)}(\mathbf{X}^{(u)}) \\ \mathbf{X}^{(u)} \end{pmatrix} \le \overline{\mathbf{R}}^{(u)}, \; q \le l = \sum_{j=1}^{l} \delta_j^{(u)} \le \overline{l}, \; 1 < N < \overline{N}, \; \det(\mathbf{C}_{\alpha}) \le \overline{C} \tag{2}$$

where det(·) –a determinant of the matrix; $\mathbf{C}_{\boldsymbol{\alpha}}$ – a covariance matrix of parameters of elements; $\mathbf{R}^{(u)}$ – vector of the u-th state variables; $\underline{\mathbf{R}}^{(u)}$ and $\overline{\mathbf{R}}^{(u)}$ - lower and upper limits of variation in the state variables; u – state index; N – number of states; $\overline{N}$ – maximum admissible number of states (experiments); $\mathbf{X}_{\mathbf{R}}^{(u)}$, $\mathbf{Y}_{\mathbf{R}}^{(u)}$ – vectors of independent and dependent u-th state variables, respectively; $\boldsymbol{\delta}^{(u)}$ – $\overline{l}$-dimensional vector of Boolean variables in the u-th state with components: $\delta_j^{(u)} = 1$, if the j-th variable is measured in the u-th state, and $\delta_j^{(u)} = 0$, otherwise, $j = 1,...,\overline{l}$; $\boldsymbol{\alpha}$ – vector of parameters of elements, $\boldsymbol{\alpha} = \{\mathbf{s},\mathbf{k}\}$ ($\mathbf{s}$ is vector of hydraulic resistances coefficients; $\mathbf{k}$ – vector of thermal-physical characteristics of elements – heat transfer coefficients; $\overline{l}$ – number of parameters from a set of components of vector $\mathbf{R}$, available for measurements; q – minimum necessary number of measurements; l – number of measurable parameters; $\overline{C}$ – a set level of accuracy.

The information criterion is represented by the determinant of the covariance matrix of parameters of elements (D-criterion) as a measure of accuracy for the estimation of vector of unknown parameters on the whole.

The overall algorithm of successive active identification of pipeline systems when planning optimal tests can be represented by the following steps: 1) planning the test conditions [15]; 2) placement of measurement devices [16]; 3) experiment; 4) processing of its results [17].

For the initial model for thermal-hydraulic tests, we propose the model of thermal-hydraulic conditions [18].

In [17, 19-20], the authors demonstrate that despite the increased complexity of mathematical models of thermal-hydraulic conditions, they manage to much faster redefine (compared to the models of isothermal flow distribution), owing to the involvement of temperature measurements.

3 Case study

The case study aims to demonstrate the advantages of the joint thermal-hydraulic tests over separate ones. To this end, we set the following objectives: 1) to perform a cycle of calculations on a concrete example in order to conduct different types of tests with the use of various mathematical models to check the effect of the active identification algorithm; 2) to study the influence of the criterion choice on the planning results; 3) to investigate the impact of conditions on the optimal choice of measuring devices

The studies were conducted for a conventional pipeline system depicted in Figure 1.

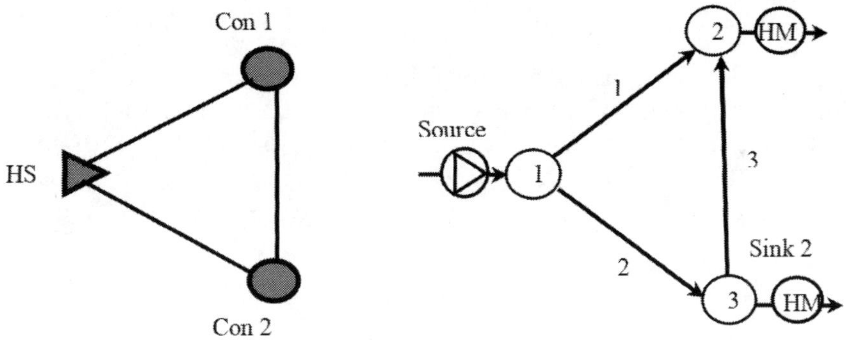

HS – heat source; Con 1, Con 2 – consumers 1 and 2, respectively; HM – heat meter.

Fig. 1. A block diagram and a calculation scheme of the conventional pipeline system.

Calculation conditions: Pressure at the first node is considered to be fixed. . For the variables that could be measured we assumed the following values of standard deviations. To plan the first experiment, it is necessary to specify the initial approximation of the resistance and heat transfer coefficients values and their standard deviation. The rule of their setting by expert is presented in [21].

To analyze the results of the numerical experiment we specified the limiting values according to the used criterion: for hydraulic parameters – $\bar{C}_{C_s} = 8.0 \cdot 10^{-27}$; and for thermophysical parameters – $\bar{C}_{C_k} = 1.0 \cdot 10^{-6}$.

The problem-solving process can be stopped only by the estimation results, therefore with the constraint imposed on the determinant of the covariance matrix of parameters of elements the question arises about how to impose this constraint and how to assess the quality of the data obtained from the tests based on the constraint. Consequently, in the process of the numerical studies, it was decided to introduce an additional criterion determined as $\gamma = \max_j \frac{\hat{\sigma}_j^2}{\tilde{\sigma}_j^2}$ (where $\hat{\sigma}_j^2$ – variance of unmeasured state variables that was obtained from the estimation; $\tilde{\sigma}_j^2$ – variance of direct measurement of the state variables).

Criterion γ demonstrates the relation between the projected variance and the variance of direct measurement. Thus, the criterion makes it possible to characterize the adequacy of the applied pipeline system models.

Planning of the state for tests. The problem is large dimensional, nonlinear and multiextremal. The authors of [15] propose a solution to this problem by the widely accepted genetic algorithms. The solution is reduced to the guided search with elements of "stochastic walk" with the help of selection operators. As the final result, some subset of points with the minimum deviation of the objective function is formed.

Planning of the state for tests makes it possible to determine an optimal composition of the vector of varied parameters, which leads to extraction of maximum information on the whole system.

Dependent variables of model [18] included vectors of: nodal flow rates, flow rates in the branches, temperatures at the beginning and end of the branches and temperatures of mixed flows at nodes and temperature of sinks $\mathbf{Y} = \left(\mathbf{Q}^{\mathrm{T}}, \mathbf{x}^{\mathrm{T}}, \mathbf{t}_b^{\mathrm{T}}, \mathbf{t}_{end}^{\mathrm{T}}, \mathbf{T}^{\mathrm{T}}\right)^{\mathrm{T}}$. The independent variables of model [18] were represented by vectors of nodal pressures, element variables and temperature of source $\mathbf{X} = \left(\mathbf{P}^{\mathrm{T}}, \boldsymbol{\alpha}^{\mathrm{T}}, \mathbf{T}_s^{\mathrm{T}}\right)^{\mathrm{T}}$. Vector of element variables was $\boldsymbol{\alpha} = \{\mathbf{s}, \mathbf{k}\}$. Full vector of measured variables consisted of vectors of nodal pressures, nodal flow rates, flow rates in the branches and vectors of source temperatures, temperatures at the beginning and end of the branches $\mathbf{Z}_1 = \left(\mathbf{P}^{\mathrm{T}}, \mathbf{Q}^{\mathrm{T}}, \mathbf{x}^{\mathrm{T}}, \mathbf{t}_b^{\mathrm{T}}, \mathbf{t}_{end}^{\mathrm{T}}, \mathbf{T}_s^{\mathrm{T}}\right)^{\mathrm{T}}$.

After the a priori information had been specified the state was designed. It was supposed that the temperature of source and sinks at consumers 1 and 2 that played a role of varying state variables could be controlled. The ranges of variation in the variables were: $290 \le Q_2 \le 490$, $10 \le Q_3 \le 210$, $70 \le T_s \le 95$. For example, some arbitrary permissible state satisfying these constraints with the parameters: $T_s = 70$, $Q_2 = 290$, $Q_3 = 10$ has $\gamma_{\max} = 1.28861$. Using the methodology of active identification proposed above, we obtained a state ($T_s = 95$, $Q_2 = 340$, $Q_3 = 60$) with $\gamma_{\max} = 0.99999675 \approx 1$. It can be seen that the criterion $\gamma_{\max}$ in obtained optimal state is 30% better than in the permissible state.

In this case, the accuracy of estimating the immeasurable parameters is commensurable with their direct measurement.

Planning of optimal placement of meters for pipeline system tests. Formation of an optimal set of meters is based on the algorithm described in [13]. The algorithm is intended to sequentially reduce a maximum admissible set of meters by removal of the measurement that would degrade the information criterion to the minimum extent. This algorithm allows the global solution to be obtained in the finite limited number of steps.

The calculation results for the planned thermal-hydraulic condition are presented in Table 1.

Table 1. Optimal placement of meters in planned condition.

RM	$\det(\mathbf{C}_\alpha)$	RM	$\det(\mathbf{C}_\alpha)$	RM	$\det(\mathbf{C}_\alpha)$	RM	$\det(\mathbf{C}_\alpha)$
Initial composition of measurement vector $P_1, P_2, P_3, Q_1, Q_2, Q_3, x_1, x_2, x_3, T_s, T_{sink2}, T_{sink3}, t_{b1}, t_{b2}, t_{b3}, t_{end1}, t_{end2}, t_{end3}$							
Q_1	$0.8165.10^{-41}$	t_{b2}	$0.0825.10^{-40}$	t_{end1}	$0.0921.10^{-40}$	t_{end2}	$0.1324.10^{-34}$
P_2	$0.9314.10^{-41}$	t_{b1}	$0.1795.10^{-39}$	x_3	$0.1692.10^{-37}$	t_{end3}	$0.6844.10^{-31}$
P_3	$0.1031.10^{-40}$	t_{b3}	$0.1516.10^{-39}$				
Optimal composition of measurement: $P_1, Q_2, Q_3, T_s, T_{sink2}, T_{sink3}, x_1$							

*RM – removed measurement.

As is seen from the Table, the conditions have a different optimal set of meters. However, the algorithm enables us to terminate the problem solution at any step and to select the variant of a set of meters with their required number and minimum loss in accuracy.

Effective of use of a thermal-hydraulic model for planning the tests of heating networks. The solution results of the problem of conditions planning for different types of tests show that in the planning of hydraulic tests the condition with the optimal criterion value does not coincide with the condition in the planning of thermal tests.

The solution results also reveal that the target accuracy level in running hydraulic tests on hydraulic condition model can be achieved using 4 conditions. The use of additional information on temperatures in the hydraulic tests on model of thermal-hydraulic conditions [18] enables a decrease in the number of planned conditions to 1 and improvement in an evaluation accuracy of hydraulic parameters.

In the thermal tests on thermal condition model, the required level of evaluation accuracy of thermophysical parameters can be achieved on the basis of 2 planned conditions.

In the joint thermal-hydraulic tests only one condition is planned. The results show that the use of $\det C_{sk}$ as an information criterion provides for a high accuracy of both hydraulic and thermophysical parameters of pipeline systems. Hence, from the viewpoint of reduction in the number of planned conditions, and improvement of evaluation accuracy of both hydraulic and thermophysical parameters of pipeline systems it is preferable to carry out joint thermal-hydraulic tests.

4 Conclusions

The research is focused on the specific features of planning of joint thermal-hydraulic conditions.

The algorithm of planning of the state for tests and optimal placement of meters is tried out using the nonisothermal flow distribution model.

Possible reduction of the experiments is exemplified numerically, which potentially leads to: 1) reduction in labor intensity; 2) improvement in accuracy against the traditional approaches.

The study demonstrates that the proposed algorithm allows the planning of tests using different flow distribution models.

References

1. RD34.20.519-97 (Service of Excellence, Moscow, 1997) [in Russian]
2. RD34.09.255-97 (Service of Excellence, Moscow, 1998) [in Russian]
3. RD 153.34.1-20.526-00 (Open Joint-Stock Company "VTI", 2001) [in Russian]
4. Srinivasa Lingireddy, Lindell E. Ormsbee, Civil Engineering and Environmental Systems, **19(1)**, 13 (2012)
5. S. Alvisi, E. Creaco, M. Franchini, Urban Water Journal, **8 (4)**, 203 (2011)
6. Guoping Yu., Roger S. Powell, Intern. Journal of Systems Science, **25 (12)**, 2155 (1994)
7. Ilyas Eker, Tolgay Kara, ISA Transactions, **42 (3)**, 461 (2003)
8. B. Coulbeck and C.H. Orr, Civil Engineering Systems, 1, 139 (1984)
9. Alireza Yazdani and Paul Jeffrey, Interdisciplinary Journal of Nonlinear Science, **21 (1)**, 11 (2011)
10. A. P. Merenkov and V. Ya. Khasilev, *The Theory of Hydraulic Circuits* (Science, Moscow, 1985) [in Russian]
11. O. A. Grebneva, Candidate's Dissertation in Engineering (ESI SB RAS, Irkutsk, 2005)
12. O.A. Grebneva, N.N. Novitsky, *Pipeline systems of power engineering*, 177 (Science, Novosibirsk, 2007) [in Russian]
13. O.A. Grebneva, N.N. Novitsky, Thermal Engineering, **61 (10)**, 754 (2014)
14. O.A. Grebneva, N.N. Novitsky, *Pipeline Energy Systems: Methodological and Applied Problems of Mathematical Modelling*, 220 (Science, Novosibirsk, 2015) [in Russian]
15. O.A. Grebneva, *System research in power engineering*, 48 (ESI SB RAS Irkutsk, 2001) [in Russian]
16. O.A. Grebneva, N.N. Novitsky, Thermal Engineering, **61 (9)**, 685 (2014)
17. N.N. Novitsky, *Estimation of the parameters of hydraulic circuits* (Science, Novosibirsk, 1998) [in Russian]
18. N.N. Novitsky, V.V. Tokarev, Thermophysics and Aeromechanics, **14 (2)**, 289 (2007) [in Russian]
19. N.N. Novitsky, Candidate's Dissertation in Engineering (ESI SB RAS, Irkutsk, 1986). [in Russian]
20. N.N. Novitsky, V.G. Sidler, Izv. of AS of USSR. Energy and transport, **4**, 155 (1984) [in Russian]
21. O.A. Grebneva, N.N. Novitsky, *Pipeline systems of power engineering*, 245 (2008) [in Russian]

Optimization of hydraulic modes of distribution heat networks by dynamic programming*

Alexandr V. Lucenko[1,†]

[1] Melentiev Energy Systems Institute of Siberian Branch of the Russian Academy of Sciences (ESI SB RAS), Pipeline Energy Systems Department, 130, Lermontov str., Irkutsk, Russia, 664033

Abstract. An original modification of the dynamic programming method is proposed, which is designed to optimize the hydraulic modes of distribution heat networks, based on special properties of the problem. It is demonstrated that the proposed modification of the dynamic programming method has a high computational efficiency in comparison with possible alternative methods of discrete-continuous optimization.

1 Introduction

Russia raises the questions of energy efficiency, and heat supply systems (HSS) have large energy-saving resources due to the non-optimal nature of their operating modes. The task of planning the operating modes of the HSS arises in preparation for the heating season. In practice, this problem is solved by multivariate calculations of the mode [1,2,**Ошибка! Источник ссылки не найден.**]. The choice of ways to organize the modes is assigned to the engineer, and the quality and optimality of decisions depends both on his experience and qualification, and on the scale and complexity of the HSS. This does not guarantee the optimality, and often the acceptability, of the developed regimes. This increases the relevance of the development and application of methods and software packages for calculating the optimal HSS. There are some factors, hindering the automation of solving this problem. These are: high dimensionality of the applied flow distribution models, their nonlinearity, several optimality criteria, discreteness of variables, etc. For these reasons, there are no methods and software packages that are suitable for practical use.

Recently there are works devoted to this problem. Most of them are devoted to TSS of small dimension, for example [4]. Genetic algorithms are used in [5], which is associated with large computational costs. Often, works devoted to the optimization of TSS regimes are of a private nature and limited scope. Thus, in [6], the optimization of the mode is performed in order to minimize the total fuel consumption, and in [7] - the total power consumed at pumping stations. Real-time management task is given enough attention, for example [6,8,9]. The tasks of reconstruction and development are devoted to a large number of works [10,11]. In all these articles, in order to overcome the dimensionality

* The research was carried out within the project III.17.4.3 of the Fundamental research program of SB RAS (AAAA-A17-117030310437-4) with finance support of RFBR and the Government of Irkutsk Region in the framework of research project № 17-48-380021

† Corresponding author: luc_alex@mail.ru

problem, the HSS calculation schemes are aggregated. This can adversely affect the accuracy and optimality of solutions.

To overcome the dimension problem ISEM SB RAS developed a multi-level approach to optimizing HSS modes [12]. It includes the following steps:1) decomposition of TSS into backbone (BHN) and distribution (DHN) heat networks; 2) finding the limits of the mode parameters at decomposition points that guarantee the existence of permissible modes of DHN; 3) optimization of the BHN mode, taking into account these limitations; 4) optimization of DHN modes taking into account the results of the previous stage. At the BHN level, a part of the heat networks, which contains all heat sources and pumping stations, is looped (in a single-line image), passive branched networks to the end users - to the DHN level. This allows you to take into account the costs associated with maintaining the mode, only when optimizing the modes of BHN [13].

In this paper, we consider the problem related to item 4 of this approach and the adaptation of the method of dynamic programming (DP) to its solution. This task is important in the development of modes for a single DHN, or HSS, which has a single source and branched passive thermal networks.

At the moment there are many implementations of the DP method used to optimize the modes and parameters of the pipeline systems tree structure [10]. For the case of tree-like pipeline systems, the conditions for the applicability of the DP are fulfilled: the additivity of the objective function and Bellman's optimality principle [14]. Sumarokov S.V. for the general case of multi-loop networks with an unknown flow distribution proposed a method of multi-loop optimization (MLO) [15], that assumes the iterative process. The main problem of applying DP to multi-circuit pipeline systems with known flow distribution is related to the need to comply with Kirchhoff's second law on contours. For this case, a modification of the MLO method [16], which assumes the iteration process, at each iteration, the pressure calculations are carried out by the DP method on a tree-like form of the calculated scheme with subsequent correction of pressure limitations. The main drawback of this approach is its iterativity and poor convergence.

2 Problem statemant

The main requirements for DHN modes are related to their permissibility and minimum labor input for their provision. The last can be reduced to minimizing the number of controls. It is advisable to reduce the overall level of pressure. This makes the task of the task of two-criteria optimization.

The main elements of the DHN are pipelines and consumers. We denote I_{PL} and I_C - sets of indexes of branches modeling elements of the first and second types so that $I_{PL} \cap I_C = \varnothing$ and $I_{PL} \cup I_C = I$ – set of indices of all branches of DHN. We will use the controlled flow distribution model [16]:

$$\mathrm{U}(\mathbf{X}) = \begin{pmatrix} \mathbf{U}_1(\mathbf{x},\mathbf{Q}) \\ \mathbf{U}_2(\mathbf{P},\mathbf{y}) \\ \mathbf{U}_3(\mathbf{x},\mathbf{y},\mathbf{z}) \end{pmatrix} = \begin{pmatrix} \mathrm{A}\mathbf{x}-\mathbf{Q} \\ \mathrm{A}^{\mathrm{T}}\mathbf{P}-\mathbf{y} \\ \mathbf{y}-\mathbf{f}(\mathbf{x},\mathbf{z}) \end{pmatrix} = \mathbf{0}, \qquad (1)$$

where A – incidence $(m \times n)$-matrix of a connected oriented graph of the calculated scheme of the HSS; n, m – number of branches and nodes of the calculated scheme; $\mathbf{Q}$ – m-dimensional vector of nodal flows; $\mathbf{P}$ – m-dimensional vector of nodal pressures; $\mathbf{x}$, $\mathbf{y}$ – n-dimensional vectors of flow and pressure differentials on the branches of the hydraulic circuit; $\mathbf{f}(\mathbf{x},\mathbf{z})$ – n-dimensional vector function with elements $f_i(x_i, z_i) = z_i s_i x_i |x_i|$ [16], $i = \overline{1,n}$; $\mathbf{z}$ – control vector; s_i – rated hydraulic resistance; $\mathbf{X}=\{\mathbf{R},\mathbf{z}\}$ – vector of real variables; $\mathbf{R}=\{\mathbf{P},\mathbf{x},\mathbf{y},\mathbf{Q}\}$ – vector of mode parameters.

As the boundary conditions for the DHN, pressures are set in the nodes of the connection between the DHN and the BHN (in the supply and return pipelines) and the flows in all other nodes. Technical and technological requirements for the permissibility and feasibility of the mode and technologically permissible limits for changing controls are reduced to the need to ensure conditions $\underline{\mathbf{X}} \le \mathbf{X} \le \overline{\mathbf{X}}$, where $\overline{\mathbf{X}}, \underline{\mathbf{X}}$ - The upper and lower constraints on the variation of the **X** components. We introduce the vector of Boolean variables $\boldsymbol{\delta}$, whose components denote the presence of control on the i-th branch of the design scheme ($i \in I_{PL}$), and inequality $\underline{z}_i \le z_i \le \underline{z}_i + \delta_i(\overline{z}_i - \underline{z}_i)$. Then the system of inequalities turns into

$$\underline{\mathbf{X}} \le \mathbf{X} \le \overline{\mathbf{X}}(\boldsymbol{\delta}). \quad (2)$$

As an optimality criterion for the number of controls, we use $F_z = \sum_{i \in I_{PL}} \delta_i$ [16]. let $F_P = \sum_{j \in J} P_j / m$ [16] be an indicator of the overall level of pressure in the network. Proceeding from practice, the task of two-criteria optimization with discrete main criterion is most expedient. We apply the principle of lexicographic ordering of criteria. Then the problem of two-criteria optimization will take the form

$$\min F_P \text{ under constraints (1), (2) and } F_z \le F_z^*. \quad (3)$$

Here F_z^* – the least number of controls on the passive branches necessary to achieve an permissible mode. Known: the topology of the calculated scheme (matrix A), boundary conditions, the coefficients of the hydraulic characteristics of the branches (s_i), the limits of the permissible change of the unknowns. $(\underline{\mathbf{X}}, \overline{\mathbf{X}})$.

3 Modified DP method

Features of DHN are: 1) tree in a single-line representation and multi-contour in a two-line; 2) the supply and return pipelines are symmetrical to each other and connected through consumers whose coolant flow rates are fixed; 3) fixed flow distribution. Due to these features, the DHN calculated scheme can be reduced to one branch using techniques of equivalent of consecutively and in parallel connected branches.

To facilitate control over the implementation of Kirchhoff's second law on contours, it is proposed to "reduce" the contours to a parallel connection of the two branches. To preserve the ability to apply DP on "reduced" contours, techniques for the equivalence of piezometric graphs of consecutively and parallel connection of branches are proposed. The direct approach of the proposed approach is to reduce the piezometric graphs of the DHN calculated scheme with the rejection of unacceptable and non-optimal variants, leaving only the optimal piezometric graph. The reverse approach consists in its restoration over all elements of the original DHN calculated scheme.

We introduce the set of indices of all nodes J and the set of branches of the calculated scheme L. Let all the branches (L_i, $i \in I$) be directed along the flow. Denote the initial and final nodes of the branch as f_i and l_i. The intervals of pressure change in the nodes will be divided into pockets (P_j^r), $j \in J$. We assign to each branch a set of segments of piezometric graphs ($L_i = \bigcup\{g_i^k\}$) starting at the initial node of the branch, and terminating in the final one. To each such segment we assign the value of the increment of the optimality criterion (ΔF_i^k). We denote the initial pocket as ϕ_i^k, final – φ_i^k.

The technique of equating parallel branches of one. If on two branches connected in parallel ($i1$, $i2$), there is a pair of segments of piezometric graphs for which $\phi_{i1}^{k1} = \phi_{i2}^{k2}$ и $\varphi_{i1}^{k1} = \varphi_{i2}^{k2}$, then such a pair becomes a segment of the piezometric graph of the equivalent branch beginning and ending in the corresponding pockets. The values of the increments of the optimality criterion are added together. The segments of the piezometric graphs, not included in any pair, are rejected.

When equating two branches ($i1$, $i2$) connected in series ($f_{i2} = l_{i1}$), it is proposed to find all such pairs of segments of piezometric graphs on these branches (g_{i1}^{k1}, g_{i2}^{k2}) that the node f_{i2} satisfies the relation $\varphi_{i1}^{k1} = \phi_{i2}^{k2}$. Of the several pairs of segments connecting the same pockets, the best remains. In the case of solving a multicriteria problem with lexicographical ordering of criteria, for example (3), after rejecting by the main criterion, it is proposed to reject pairs by secondary criteria. Each of the pairs found is transformed into a section of the piezometric graph with the corresponding initial and final pockets. The increments in the optimality criterion are added together.

To restore the optimal trajectory of the phase variables, it is necessary for each equivalent branch to remember which fragment it is equivalent to, and for each equivalent path of the phase variable, which segments it equals.

Unlike the MLO method, the proposed method does not imply iterative processes. The advantage of the proposed method over methods based on the method of internal points [16] is the linear growth in DHN dimension of the necessary computational costs.

4 Computational experiments

To test the efficiency of the proposed method, a series of computational experiments was conducted. Here is a typical example. For a test RTS, the scheme of which in a single-line image is shown in Fig. 1-a, the problem (3) was solved. In Fig. 1-b shows its calculated scheme. Branches 8 to 11 are consumers. Coolant flow rates are set on them x_8=100, x_9=50, x_{10}=100, x_{11}=150. The minimum pressure drop across these branches is 15. The fluxes in nodes 2 to 15 are zero. P_1=100 water meter, P_{16}=30. The pressure in node 9 was limited from below by a value of 70 water meter, in node 6 - from above with a value of 60. The resistance vector is (0,00003125, 0,0005, 0,00011111, 0,0005, 0,0004, 0,0003, 0,0002222, 0,0001, 0,0001, 0,0001, 0,0001, 0,0005, 0,0004, 0,0003, 0,0002222, 0,0005, 0,00011111, 0,00003125).

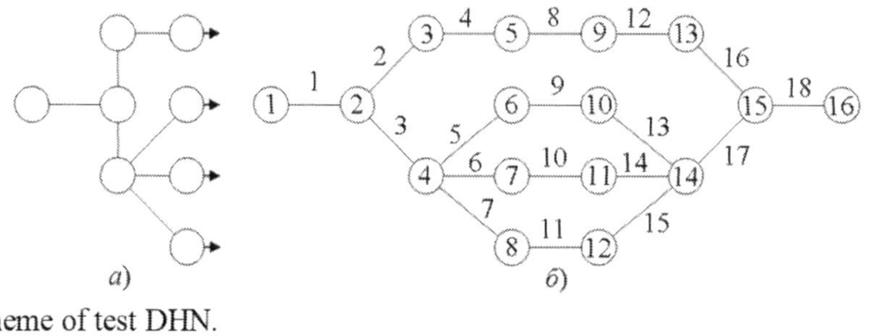

Fig. 1. Scheme of test DHN.

To achieve the permissibility of the regime, it was necessary to install throttling devices on branches 5 and 12. The pressure vector found is equal to (100, 95, 90, 85, 85, 60, 82, 80, 70, 45, 47, 79, 40, 44, 35, 30) water meter. The verification of the results was carried out by the majorizing sequence method [16]. The verification showed the correctness of the results of the proposed method. Also, the running time of the proposed method was significantly shorter than the running time of the majorizing sequence method.

Computational experiments showed the efficiency of the proposed method.

5 Conclusions

1. The mathematical formulation of the discrete-continuous problem of two-criteria optimization of the DHN hydraulic regimes arising at the stage of regime planning is given. Currently, there are no special methods for its solution, suitable for practical use.

2. To solve this problem, an original method is proposed, consisting in folding with the rejection of unrealizable and suboptimal piezometric graphs of the DHN calculation scheme into a piezometric graph of one branch with its subsequent unfolding on the entire DHN.

3. It is demonstrated that the proposed method allows solving multicriterion optimization problems.

4. The proposed method is implemented in the form of a research program. The results of its application for numerical calculations illustrating its effectiveness are presented.

References

1. H. Boysen, J.E. Thorsen, EHP **4** (2007)
2. F. Wernstedt, P. Davidsson, *Proc. of the First Intern. Conf. on Agreement Technologies, Dubrovnik, Croatia, October 15-16, 2012,* p. 423-437. (2012)
3. N.N. Novitsky, M.G. Sukharev and others, *Pipeline energy systems. Methodical and applied problems of mathematical modeling.* (2015)
4. D. Buoro, P. Pinamonti, M. Reini, App. Energy, **124**, (2014)
5. A. Sciacovelli, E. Guelpa, V. Verda, *Proc. of Intern. Mechanical Engineering Congr. IMECE2013: 2013 ASME & Exposition San Diego, California, November 15-21.*
6. X.S. Jiang, Z.X. Jing, Y.Z. Li, Q.H. Wu, W.H. Tang, Energy, **64**, (2014.)
7. S. Cosentino, E. Guelpa, R. Melli, A. Sciacovelli, E. Sciubba, C. Toro, V. Verda, *Proc. of the Intern. Mechanical Engineering Congr. And Exposition «ASME» 2014 November 14-20. Montreal, Quebec, Canada.* (2014)
8. Z.X. Jing, X.S. Jiang, Q.H. Wu, W.H. Tang, B. Hua, Energy, **73**, (2014.)
9. T. Laajalehto, M. Kuosa, T. Makila, M. Lampinen, R. Lahdelma, Appl. Therm. Eng. **69**, (2014)
10. A.P. Merenkov, V.Y. Hasilev, *Theory of hydraulic circuits.* (1985)
11. C. Haikarainen, F. Pettersson, H. Saxen, Chem. eng. trans. **35**, (2013)
12. A.V. Lucenko, N.N. Novitsky, *Methodological issues of reliability research for large energy systems.* Issue. 64. (2014)
13. A.V. Lucenko, N.N. Novitsky, J. Global Optim., **66**, 1 (2016)
14. R.E. Bellman, *Dynamic Programming.* (1957).
15. S.V. Sumarokov, A.V. Hramov, Methods of optim. oper. research (appl. math.). (1976)
16. A.P. Merenkov, E.V. Sennova, S.V. Sumarokov, et al. *Mathematical modeling and optimization of heat, water, oil and gas supply systems.* (1992)
17. A.V. Lutsenko, N.N. Novitsky, Sci. Bull. NSTU. **64**, 3. (2016)

Quantitative assessment of the identifiability of pipeline systems*

Nikolay N. Novitsky[1], *and Oksana* A. Grebneva[1,†]
[1] Melentiev Energy Systems Institute of Siberian Branch of the Russian Academy of Sciences (ESI SB RAS), Pipeline Energy Systems Department, 130, Lermontov str., Irkutsk, Russia, 664033

Abstract. The article is devoted to the issues of quantitative assessment of the identifiability of the pipeline systems (heat, water, gas supply systems etc.). Identifiability is first considered as a complex property, including such particular properties as observability and parametric identifiability. A brief description of the topic relevance and a review of available development in this sphere allow giving the structuring of identifiability analysis problems. The technique of differentiate quantitative analysis of this property is disclosed. It based on the use of analytical expressions for covariance matrices of parameters. New concepts of experimental matrices, parametric identifiability and observability of pipeline systems are introduced. Analytic expressions for these matrices are given. The substantiation of the integral indicators of the pipeline systems identifiability is presented, including the covariance matrix determinant for the estimated parameters and the relative variance of the prediction for non-measurable state parameters. The analytical interrelation of these indicators is opened. These indicators can be accepted in a role of criteria at decision of synthesis problems for optimal measurements composition.

1 Introduction

The urgency of the article theme is due to increasing rates of technological transformation, renovation and modernization of pipeline systems (PLS) due to the integration of new equipment, devices of control, monitoring and measuring, computer technology, mathematical modelling methods, computer systems for measuring and processing measurements and making control decisions. The main purpose of these transformations has recently been increasingly associated with the intelligent automation of PLS. First of all, the intelligence of PLS is the intelligent automation of control. In this regard, there are questions of quantitative estimation of the intellectual property of the PLS from the point of view of their satisfaction with the basic cybernetic properties – controllability and identifiability. As [1,2] were shown the main attention is paid to the identification of PLS as a complex property, including such particular properties as observability and parametric identifiability.

* The research was carried out within the project III.17.4.3 of the Fundamental research program of SB RAS (AAAA-A17-117030310437-4)
† Corresponding author: oksana@isem.irk.ru

2 The problem of pipeline systems identifiability

Identifiability is understood as the possibility of restoring the mathematical model of PLS with accuracy to parameters from the results of measurements and observations. The relevance of independent study of this property in respect to PLS is determined by the following main circumstances: 1) the characteristics of PLS equipment undergo significant changes during operation, and the coefficients of these characteristics are inaccessible to direct measurement or observation; 2) lack of information on the actual characteristics of PLS reduces the effectiveness and validity of decisions on control, maintenance and repair, reconstruction and development of PLS; 3) the existing practice of PLS equipping with measuring devices at the main facilities (at sources, pumping stations, etc.) leads to the fact that the flow distribution processes remain unobservable, and the extent of achievement of control objectives remains uncontrolled; 4) the lack of adequate PLS models invalidates practically the entire arsenal of mathematical modelling and optimization methods for PLS control.

Recently, in connection with the emergence of advanced information and measurement systems, more attention is being paid to the development and application of identification (calibration) methods of PLS [3], water [4,5], heat [6,7], gas [8], oil supply systems [9], etc., as well as problems of optimizing of measurements composition for the possibility of using these methods [10].

In ESI SB RAS, the issues of PLS identifiability have been being systematically begun to be investigated since the early 1990th [11 - 13, etc.]. This investigation included the development of a new approach to the identification of PLS [14, 15]. In the framework of this approach, practically all methodological limitations were removed. These limitations were related to the type of model, possible combinations of measured or sought-for parameters, and other factors. Accordingly, it became possible to objectively study of the identifiability as a property of PLS itself, rather than the methods for identifying of them.

The structuring of the problems for identifiability analysis can be performed depending on [1,2]: 1) identification purposes (states, parameters, models) – analysis of observability, parametric or structural identifiability; 2) factors determining the possibility of achievement these goals – *qualitative and quantitative analysis*; 3) the conditions for studying of these factors – *a priori or a posteriori analysis* (before the experiment or after processing its results); 4) the purposes of further use of the analysis results – *differential or integral analysis*.

Thus, identifiability is a complex property that requires independent study, development of its own methodical apparatus. Below, problems and methods of a priori quantitative assessment of identifiability from the point of view of information and objective aspects of this property are considered.

3 Differential assessments of identifiability

Qualitative analysis of identifiability makes it possible to establish the principle solvability of the identification problem depending on the composition of the measured parameters [11, 12], but does not answer the question of the identification accuracy under conditions of measurement errors presence. In [14], the technique of differential quantitative analysis of identifiability was proposed. It based on the use of analytic expressions for covariance matrices (CM) of an arbitrary subset of PLS parameters.

The following main results were obtained here. Suppose that the system of equations is used as a model of PLS, which we represent in the form

$$U(\mathbf{Z}) = 0, \qquad (1)$$

where: $\mathbf{Z}$ – vector of model parameters.

We introduce the following decomposition methods for $\mathbf{Z}$:

1) $\mathbf{Z}=\{\mathbf{R},\boldsymbol{\alpha}\}$, where $\mathbf{R}$ – vector of state parameters (pressure, flow rates, etc.), $\boldsymbol{\alpha}$ – vector of the parameters of the elements (the model coefficients) specifying the internal technical condition of the PLS (the control vector u is considered here as known, deterministic given);

2) $\mathbf{Z}=\{\mathbf{Z}_1,\mathbf{Z}_2\}$, where $\mathbf{Z}_1,\mathbf{Z}_2$ – vectors of measured and non-measurable parameters;

3) $\mathbf{Z}=\{\mathbf{X},\mathbf{Y}\}$, where $\mathbf{X},\mathbf{Y}$ – vectors of independent and dependent parameters, and $\mathbf{X}$ always must provide a single-valued computation $\mathbf{Y}$ from (1);

4) $\mathbf{X}=\{\mathbf{X}_R,\boldsymbol{\alpha}\}$. That is, we assume that $\mathbf{X}$ always includes the vector of elements parameters $\boldsymbol{\alpha}$, and $\mathbf{X}_R$ is the vector of the independent state parameters. Wherein

$$\mathbf{Y}=\mathbf{Y}(\mathbf{X})=\mathbf{Y}(\mathbf{X}_R,\boldsymbol{\alpha}) \tag{2}$$

If we involve the measurement results for N conditions, we assume that $\mathbf{Z}=\{\mathbf{R},\boldsymbol{\alpha}\}=\{\mathbf{R}^{(1)},\mathbf{R}^{(2)},\ldots,\mathbf{R}^{(N)},\boldsymbol{\alpha}\}$, $\mathbf{R}^{(t)}=\{\mathbf{X}_R^{(t)},\mathbf{Y}^{(t)}\}$, $u=\overline{1,N}$,

$\mathbf{X}=\{\mathbf{X}_R^{(1)},\mathbf{X}_R^{(2)},\cdots,\mathbf{X}_R^{(N)},\boldsymbol{\alpha}\}$, $\mathbf{Y}=\{\mathbf{Y}^{(1)},\mathbf{Y}^{(2)},\cdots,\mathbf{Y}^{(N)}\}$.

The error in the definition $\mathbf{Z}$ is determined as $\boldsymbol{\varepsilon}_Z=\hat{\mathbf{Z}}-\overline{\mathbf{Z}}$, where $\hat{\mathbf{Z}}$ is some estimate of the true value $\overline{\mathbf{Z}}$ of the model parameters vector $\mathbf{Z}=\{\mathbf{R}^{(1)},\mathbf{R}^{(2)},\ldots,\mathbf{R}^{(N)},\boldsymbol{\alpha}\}$. As the value $\overline{\mathbf{Z}}$ is unknown, the error level can be determined statistically in the form of a CM $\mathbf{C}_Z\equiv\mathbf{E}(\boldsymbol{\varepsilon}_Z\boldsymbol{\varepsilon}_Z^T)$, on the diagonal of which there are component variances $\boldsymbol{\varepsilon}_Z$, and off-diagonal elements correspond to their covariance. As shown in [14], *the CM of model parameters* have a structure $\mathbf{C}_Z=\left[\begin{array}{c|c}\mathbf{C}_X & \mathbf{C}_{XY}\\ \hline \mathbf{C}_{XY}^T & \mathbf{C}_Y\end{array}\right]=\left[\begin{array}{c|c}\mathbf{C}_X & \mathbf{C}_X(\mathbf{Y}'_X)^T\\ \hline \mathbf{Y}'_X\mathbf{C}_X^T & \mathbf{Y}'_X\mathbf{C}_X(\mathbf{Y}'_X)^T\end{array}\right]$.

Here $\mathbf{Y}'_X=\partial\mathbf{Y}/\partial\mathbf{X}$ – a matrix of partial derivatives, $\mathbf{C}_X=(\mathbf{J}^T\mathbf{C}_{Z1}^{-1}\mathbf{J})^{-1}$ – a CM of estimation errors for $\mathbf{X}$, $\mathbf{C}_{Z1}$ – a CM of measurement errors of the vector $\mathbf{Z}_1$, $\mathbf{J}=\partial\mathbf{Z}_1/\partial\mathbf{X}=\mathbf{I}_{Z1}\begin{pmatrix}\mathbf{E}\\ \mathbf{Y}'_X\end{pmatrix}=\mathbf{J}(\mathbf{Y}'_X)$ – a matrix of the partial derivatives of the measured parameters as function of the independent, $\mathbf{E}$ – the identity matrix, $\mathbf{I}_{Z1}$ – a matrix of the correspondence of components of the vectors $\mathbf{Z}_1$ and $\mathbf{Z}$, consisting of zeros and ones.

Thus, $\mathbf{C}_Z$ is completely defined if $\mathbf{C}_X$ and $\mathbf{Y}'_X$ are given, and $\mathbf{C}_X$ if the matrices $\mathbf{I}_{Z1}$, $\mathbf{C}_{Z1}$ and $\mathbf{Y}'_X$ are given. Accordingly, the accuracy of the estimates $\hat{\mathbf{Z}}$ (and hence the identifiability) is determined by three factors are the composition of measurements, their accuracy and sensitivity $\mathbf{X}$ to $\mathbf{Z}_1$, determined by the implicit function $\mathbf{Z}_1(\mathbf{X})$ (PLS model) in the conditions for which these measurements are made. I.e, $\mathbf{C}_X=\mathbf{C}_X(\mathbf{I}_{Z1},\mathbf{C}_{Z1},\mathbf{X})$.

4 Private information and covariance matrices

We call the matrix $\mathbf{F}_X=\mathbf{J}^T\mathbf{C}_{Z1}^{-1}\mathbf{J}$ *the information matrix of the experiment*, as it includes all the information extracted from this experiment and is necessary for estimate the residual degree of uncertainty about the unknown parameters, regardless of the purpose of the experiment. Then $\mathbf{C}_X=\mathbf{F}_X^{-1}$ is a *CM of experiment*. Taking into account the introduced decomposition $\mathbf{X}=\{\mathbf{X}_R,\boldsymbol{\alpha}\}$, these matrices have the structure

$$\mathbf{F}_X = \mathbf{J}^T\mathbf{C}_{Z1}^{-1}\mathbf{J} = \left[\begin{array}{c|c} \mathbf{J}_R^T\mathbf{C}_{R1}^{-1}\mathbf{J}_R & \mathbf{J}_R^T\mathbf{C}_{R1}^{-1}\mathbf{J}_\alpha \\ \hline \mathbf{J}_\alpha^T\mathbf{C}_{R1}^{-1}\mathbf{J}_R & \mathbf{J}_\alpha^T\mathbf{C}_{R1}^{-1}\mathbf{J}_\alpha + \mathbf{I}_{\alpha1}^T\mathbf{C}_{\alpha1}^{-1}\mathbf{I}_{\alpha1} \end{array}\right], \quad (3)$$

$$\mathbf{C}_X = \mathbf{F}_X^{-1} = \left[\begin{array}{c|c} \mathbf{C}_{XR} & \mathbf{C}_{XR\alpha} \\ \hline \mathbf{C}_{XR\alpha}^T & \mathbf{C}_\alpha \end{array}\right], \quad (4)$$

where the following notation is used: $\mathbf{J}_R = \partial\mathbf{R}_1/\partial\mathbf{X}_R$; $\mathbf{J}_\alpha = \partial\mathbf{R}_1/\partial\boldsymbol{\alpha}$; $\mathbf{I}_{\alpha1}$ – a matrix, similar in meaning $\mathbf{I}_{Z1}$, but with respect to given pseudo-measurements $\boldsymbol{\alpha}_1$; vectors of measurement of the state parameters $\mathbf{R}_1$ and elements parameters $\boldsymbol{\alpha}_1$ are assumed to be uncorrelated. Wherein $\mathbf{J} = \begin{bmatrix} \mathbf{J}_R & \mathbf{J}_\alpha \\ \mathbf{0} & \mathbf{I}_{\alpha1} \end{bmatrix}$, $\mathbf{C}_{Z1} = \begin{bmatrix} \mathbf{C}_{R1} & \mathbf{0} \\ \mathbf{0} & \mathbf{C}_{\alpha1} \end{bmatrix}$, $\mathbf{I}_{Z1} = \begin{bmatrix} \mathbf{I}_{R1} & \mathbf{0} \\ \mathbf{0} & \mathbf{I}_{\alpha1} \end{bmatrix}$.

We apply to $\mathbf{F}_X$ the inversion rule of the block matrix [16], as a result of which we obtain finite expressions for the CM of the independent state parameters $\mathbf{C}_{XR} = \mathbf{F}_{XR}^{-1}$ and the CM of elements parameters, where

$$\mathbf{F}_{XR} = \mathbf{J}_R^T\mathbf{H}_\alpha\mathbf{J}_R, \quad (5)$$

$$\mathbf{F}_\alpha = \mathbf{J}_\alpha^T\mathbf{H}_{R1}\mathbf{J}_\alpha + \mathbf{I}_{\alpha1}^T\mathbf{C}_{\alpha1}^{-1}\mathbf{I}_{\alpha1}, \quad (6)$$

and

$$\mathbf{H}_\alpha = \mathbf{C}_{R1}^{-1} - \mathbf{C}_{R1}^{-1}\mathbf{J}_\alpha(\mathbf{J}_\alpha^T\mathbf{C}_{R1}^{-1}\mathbf{J}_\alpha + \mathbf{I}_\alpha^T\mathbf{C}_{\alpha1}^{-1}\mathbf{I}_\alpha)^{-1}\mathbf{J}_\alpha^T\mathbf{C}_{R1}^{-1}, \quad (7)$$

$$\mathbf{H}_{R1} = \mathbf{C}_{R1}^{-1} - \mathbf{C}_{R1}^{-1}\mathbf{J}_R(\mathbf{J}_R^T\mathbf{C}_{R1}^{-1}\mathbf{J}_R)^{-1}\mathbf{J}_R^T\mathbf{C}_{R1}^{-1}. \quad (8)$$

Accordingly, we call: $\mathbf{F}_{XR}$ – *a matrix of observability*, $\mathbf{F}_\alpha$ – *a matrix of identifiability*.

Note that if the vector $\boldsymbol{\alpha}$ is known and defined deterministically, then $\boldsymbol{\alpha}_1 = \boldsymbol{\alpha}$, $\mathbf{C}_{\alpha1} = 0$, $\mathbf{H}_\alpha = 0$ and the observability matrix takes the traditional form

$$\mathbf{F}_{XR} = \mathbf{J}_R^T\mathbf{C}_{R1}^{-1}\mathbf{J}_R. \quad (9)$$

Therefore, (5) gives a formula for the observability matrix, taking into account the uncertainty in the values of the vector α taken into account by $\mathbf{H}_\alpha$.

Relation (6) reveals the contribution to the resulting identifiability of a priori information about the parameters α (the second term on the right side), taking into account the uncertainty about the states parameters taken into account through the matrix $\mathbf{H}_{R1}$.

An important consequence of the introduced concepts and formulas is that: 1) if the system is identifiable, then it is observable, but not vice versa, therefore observability is a particular property of identifiability; 2) the obtained expressions for CM allow to perform a differentiate analysis of accuracy for any model parameter or a set of them, including by building confidential intervals or region with a given probability covering the true values; 3) such analysis can be performed depending on the purpose of the experiment (observation or parametric identification); 4) this technique can be used both for a posteriori and for a priori differential quantitative analysis of identifiability. The only difference is that the derivatives $\mathbf{Y}'_X$ are taken either at the point of the estimation problem solution, or at the point of some assumed values of the vector $\mathbf{X}$.

5 Integral indicators of identifiability

Differentiate analysis does not allow unambiguous comparison of different variants of experimental conditions, as each variant may be better than the other in its group of estimated parameters. For this comparison, as well as for a purposeful impact on these conditions (the identifiability synthesis), it is proposed to use integral quantitative indicators of the identifiability of PLSs as a whole [1,2,13,15]. In this case, the use of the *D*-criterion (the determinant of the CM of the estimated parameters) is expedient in the role of a basic indicator of identifiability. This criterion has the following important properties:

1) the reduction in the degree of uncertainty of the estimated PLS parameters can be related to the requirement to reduce the volume of the confidence ellipsoid, that is equivalent to the requirement of *D*-criterion minimizing [16];

2) the useful property of the *D*-criterion is that if the accuracy of a subset is of particular interest from the whole set of parameters, then the determinant of the submatrix can be applied. This submatrix is obtained by deleting rows and columns corresponding to parameters that are not of interest [16];

3) the most important property of this criterion in the context under consideration follows from the well-known result of the theory of experimental design [16,17], according to which the optimal value of the *D*-criterion corresponds to the minimum value of the maximum variance of the response (the dependent model variables);

4) it is also possible to interpret this criterion from the standpoint of Shannon information theory [1,18], as the minimum of entropy coincides with the minimum of the *D*-criterion.

Thus, for the integral estimation of observability and parametric identifiability of PLS, it is proposed using indicators:

$$D_{XR} = \det(\mathbf{C}_{XR}), \quad D_{\alpha} = \det(\mathbf{C}_{\alpha}). \tag{10}$$

The smaller values of these indicators, the less uncertainty in the corresponding groups of model parameters, and, accordingly, the higher the degree of observability and parametric identifiability of PLSs. These indicators already allow complex comparison of the experimental conditions. However their numerical values depend on the dimensionality of the variables and do not allow them to be given a physical meaning. Therefore, in the role of an auxiliary indicator, it is proposed to use the criterion of relative variance of the prediction

$$d = \max_{i \in I_2} \left(\frac{\hat{\sigma}_i^2}{\sigma_i^2} \right), \tag{11}$$

where I_2 – a set of unmeasurable state parameters, σ_i^2 – the a priori dispersion of the direct measurement of this parameter, determined by the metrological characteristics of the corresponding measuring devices, $\hat{\sigma}_i^2$ – the a posteriori (predicted) dispersion of this parameter by the model. An a posteriori estimate of the CM of non-measurable parameters is defined as

$$\mathbf{C}_{Z2} = \mathbf{J}_{Z2}\mathbf{C}_X\mathbf{J}_{Z2}^T, \text{ where } \mathbf{J}_{Z2} = \partial \mathbf{Z}_2 / \partial \mathbf{X} = \mathbf{I}_{Z2} \begin{pmatrix} \mathbf{E} \\ \mathbf{Y}'_X \end{pmatrix}, \tag{12}$$

where $\mathbf{I}_{Z2}$ – a matrix of correspondence of the components of the vectors $\mathbf{Z}_2$ and $\mathbf{Z}$, consisting of zeros and ones. From here

$$\hat{\sigma}_i^2 = (\mathbf{J}_{Z2})_i \mathbf{C}_X (\mathbf{J}_{Z2})_i^T, \; i \in I_2, \tag{13}$$

where $(\mathbf{J}_{Z2})_i$ – *i*-th row of the matrix $\mathbf{J}_{Z2}$.

Thus, the criterion (11) already makes sense of the relative accuracy of the non-measurable parameters determination in comparison with their direct measurement. If, $d \gg 1$ the identifiability is poor, and at $d \approx 1$ – high.

We will reveal the relationship between the criteria *D* and *d* for the simplest case of the analysis of observability under deterministic given α and diagonal matrix $\mathbf{C}_{R1} = diag\{\sigma_1^2, \sigma_2^2, ..., \sigma_l^2\}$, where $l = |I_1|$, I_1 – the set of measured state parameters. Here, the observability matrix (9) can be represented in the form

$$\mathbf{\Phi}_{XR} = \mathbf{J}_R^T \mathbf{C}_{R1}^{-1} \mathbf{J}_R = \sum_{i \in I_1} (\mathbf{J}_R)_i^T \sigma_i^{-2} (\mathbf{J}_R)_i \,. \qquad (14)$$

If we add some measure, for example, with an index $i = l+1$ (moving it from the set I_2 to I_1), instead of the matrix (14), which we denote by $F_{XR}(l)$, we get the observability matrix $\mathbf{F}_{XR}(l+1) = \mathbf{F}_{XR}(l) + (\mathbf{J}_R)_{l+1}^T \sigma_{l+1}^{-2} (\mathbf{J}_R)_{l+1}$.

It is known [19, 20] that for some square block matrix $\mathbf{A} = \begin{bmatrix} \mathbf{A}_{11} & \mathbf{A}_{12} \\ \mathbf{A}_{21} & \mathbf{A}_{22} \end{bmatrix}$, when $\det(\mathbf{A}_{11}) \neq 0$ and $\det(\mathbf{A}_{22}) \neq 0$, we have the relation $\det(\mathbf{A}) = \det(\mathbf{A}_{11}) \det(\mathbf{A}_{22} - \mathbf{A}_{21}\mathbf{A}_{11}^{-1}\mathbf{A}_{12}) = \det(\mathbf{A}_{22}) \det(\mathbf{A}_{11} - \mathbf{A}_{12}\mathbf{A}_{22}^{-1}\mathbf{A}_{21})$.

We denote $\mathbf{A}_{11} = \mathbf{F}_{XR}(l)$, $\mathbf{A}_{22} = -\sigma_{l+1}^2$, $\mathbf{A}_{12} = (\mathbf{J}_R)_{l+1}^T$, $\mathbf{A}_{21} = (\mathbf{J}_R)_{l+1}$. Then $\det[\mathbf{F}_{XR}(l)] \det[\sigma_{l+1}^{-2} + (\mathbf{J}_R)_{l+1} \mathbf{F}_{XR}^{-1}(l) (\mathbf{J}_R)_{l+1}^T] = \det(\sigma_i^{-2}) \det[\mathbf{F}_{XR}(l) + (\mathbf{J}_R)_{l+1} \sigma_{l+1}^2 (\mathbf{J}_R)_{l+1}^T]$. As $(\mathbf{J}_R)_i \mathbf{F}_{XR}^{-1}(l) (\mathbf{J}_R)_i^T = (\mathbf{J}_R)_i \mathbf{C}_{XR} (\mathbf{J}_R)_i^T = \hat{\sigma}_i^2$, $\det(\sigma_i^2) = \sigma_i^2$, $\det[\mathbf{F}_{XR}(l) + (\mathbf{J}_R)_{l+1} \sigma_{l+1}^2 (\mathbf{J}_R)_{l+1}^T] = \det[\mathbf{F}_{XR}(l+1)]$ и $\det(\sigma_{l+1}^2 + \hat{\sigma}_{l+1}^2) = \sigma_{l+1}^2 + \hat{\sigma}_{l+1}^2$, then

$$\det[\mathbf{F}_{XR}(l+1)] = \det[\mathbf{F}_{XR}(l)] \left(1 + \frac{\hat{\sigma}_{l+1}^2}{\sigma_{l+1}^2} \right). \qquad (15)$$

Thus, the determinant of the observability matrix increases by an amount $1 + d_{l+1}$ with adding a measurement $l+1$. As $\det(\mathbf{F}_{XR}) = 1/\det(\mathbf{F}_{XR}^{-1}) = 1/\det(\mathbf{C}_{XR}) = 1/D_{XR}$, then the desired interrelation of the exponents d and D has the form

$$D_{XR}(l+1) = D_{XR}(l) / \left(1 + d_{l+1}\right). \qquad (16)$$

It is seen that when adding a measurement, the indicator D_{XR} decreases inversely proportional to the value $1 + d_{l+1}$.

On these principles, algorithms for measurements placement [13, 15, etc.] have been developed. These algorithms provide the possibility of successively improving of the measurements composition by the *D*-criterion, being guided by the current values of the d-criterion. Possible formulation of problems for optimization of the measurements composition are minimization of the measurements number, taking into account limitations on the accuracy of identification, or maximization of accuracy with a limitation on the measurements number. The corresponding algorithms allow finding the optimal solution for the final steps; take into account the permissible locations of new measuring devices and the existence of existing ones.

6 Conclusion

1. Due to the commonality of development and functioning problems for various types and purposes PLSs, it is justified to develop complex intellectuality indicators of PLS from the point of view of their satisfaction with the basic cybernetic properties – controllability and identifiability.
2. The relationship between the parametric identifiability and the observability of PLS is revealed. It is shown that observability is a particular property of identifiability under conditions of uncertainty of actual PLS characteristics.
3. Integral quantitative indices of PLS identifiability and methods of their calculation are proposed. Also it is proposed *D*-criterion, which can give an interpretation of information amount by Shannon and d-criterion that reflects the predictive properties of models obtained as a result of identification. The interrelation of these criteria is revealed.

4. The proposed criteria can potentially be used to optimize information and measurement systems, as well as to develop standards for the identifiability of PLSs.

References

1. N.N. Novitsky, *Pipeline Energy Systems: Methodological and Applied Problems of Mathematical Modelling*, 159 (Science, Novosibirsk, 2015) [in Russian]
2. N.N. Novitsky, *System research in power engineering: a retrospective of scientific trends*, 279 (Science, Novosibirsk, 2010) [in Russian]
3. K. Menberg, Y. Heo, R. Choudhary, Proceed. of the 15th IBPSA Conf., 1212 (2017)
4. Y. Zheng, T. Walski, R. Mankowski, G. Herrin, R. Gurrieri, M. Tryby. AWWA IMTech Conference, 1 (2002)
5. S. Lingireddy, L. E. Ormsbee, Civil. Eng. and Env. Syst., **19(1)**, 13 (2000)
6. J. Wang, Z. Zhou, J. Zhao, Energy Conversion and Management, **120**, 294 (2016)
7. V.I. Panferov. Bulletin of SUSU, **13**, 87 (2005)) [in Russian]
8. M.G. Sukharev, R.V. Samoylov, *Analysis and control of stationary and non-stationary conditions of gas transport* (Russian State University of Oil and Gas, Moscow, 2016) [in Russian]
9. A.F. Barkhatov, D.A. Terre, Oil and gas business, **3**, 282 (2015) [in Russian]
10. L. Cozzolino, R. Morte, A. Palumbo, D. Pianese, Civil Engineering and Environmental Systems, **28:1**, 75 (2011)
11. N.N. Novitsky, *Methods of analysis and optimal synthesis of pipeline systems*, 142 (ESI SB RAS, Irkutsk, 1991) [in Russian]
12. N.N. Novitsky, *Mathematical modeling and optimization of heat, water, oil and gas supply systems*, 215 (Science, Novosibirsk, 1992) [in Russian]
13. N.N. Novitsky, O.A. Grebneva, *Pipeline energy systems: models, applications, information technologies*, 90 (Oil and gas, Moscow, 2000) [in Russian]
14. N.N. Novitsky, *Estimation of hydraulic circuit parameters* (Science, Novosibirsk, 1998) [in Russian]
15. N.N. Novitsky, Proceedings of the Russian Academy of Sciences. Power Engineering, **6**, 87 (2000) [in Russian]
16. V.V. Fedorov, *Theory of Optimal Experiment* (Science, Moscow, 1971) [in Russian]
17. J. Kiefer, J. Wolfowitz, The Equivalence to Two Extremum Problems. Canad. J. Math., **12**, 363 (1960)
18. E. Shannon, W. Weaver, *The Mathematical Theory of Communication* (Univ of Illinois Press, 1963)
19. F.R. Gantmakher, *Theory of matrices* (Science, Moscow, 1988) [in Russian]
20. R. A. Horn, C. R. Johnson, *Matrix Analysis* (Cambridge University Press, Cambridge, 2013)

Regression model for heat consumption monitoring and forecasting

Tatyana Dobrovolskaya[1*], and *Valery* Stennikov[1]

[1]Melentiev Energy Systems Institute of Siberian Branch of the Russian Academy of Sciences (ESI SB RAS), Pipeline Energy Systems Department, 130, Lermontov str., Irkutsk, Russia, 664033

Abstract. Heat supply is socially and economically important in our country. In this regard, high-quality monitoring and planning of the development of heat supply systems are a strategic vector of scientific research. This paper is focused on the studies demonstrating how to choose a methodological approach to describe changes in heat consumption in the retrospective. The change in heat consumption is described using multiple regression models. In the first part of the paper, the parameters for the regression model are determined and a statistical analysis of the obtained model is performed. In the second part of the paper, to eliminate the multicollinearity of the regression equation, the number of dependent variables in the model is reduced. A statistical analysis of the new regression model and the exponential regression model are carried out. The heat consumption values obtained using these models are compared with the statistical data. The conclusions about the quality of the obtained regression models are made. In the third part of the article, we make a forecast of heat consumption in the medium term by using a linear regression model and an exponential model.

1 Introduction

This paper is a continuation of the research on heat consumption monitoring and forecasting [1]. It is devoted to the generalization and analysis of data on energy consumption and economic indices, which are represented by time series. The studies were carried out to determine the closeness and nature of the relationship between the selected indices and the type of heat consumption dependence on the selected variables.

An analysis of the literature dealing with the prediction of energy consumption [2,3] indicates that the energy consumption is most often calculated using specific indicators for planning. However, the average heat consumption in residential buildings often does not reflect the real picture, as it does not take into account the climatic features of the region, the actual state of building envelopes, quantitative and qualitative characteristics of energy-saving measures. In industry, the forecast is based on indicators of energy intensity of various industries, that are adopted based on the foreign experience. In this case, it is quite difficult to draw analogies due to significant differences in economic development and technological potential of the countries. The researchers from other countries often present

* Corresponding author: makarova@isem.irk.ru

the studies on the use of regression models to analyse the current state and predict the levels of energy consumption [4-10]. They make an emphasis on the forecasts of electricity consumption for a short period of time (some hours or days).

This paper presents a research into a methodological approach to describe a change in heat consumption for a retrospective-calculated period. The coefficients of the multiple regression model of heat consumption are determined. A comparison of linear and nonlinear regression models of the heat consumption is performed. The research was conducted at the highest hierarchical level. The total heat consumption of the country was taken as the research object.

2 Multiple regression models of heat consumption

2.1 Determination of variables for regression model

A regression analysis is one of the most widely used statistical tools to describe the variations in a dependent variable (annual heat consumption) with independent variables used as inputs for the functions. The main goal of the regression analysis is to find an appropriate mathematical model and determine the best coefficients of the model from the given data. The major objective of the study was to identify the input parameters for the models that would describe the heat consumption in the best way.

In the first stage, we selected the parameters for the multiple regression model. A specific feature of regression models is the need to have prospective estimates of regression equations (independent variables). Consequently, it is much more difficult to choose the parameters as independent variables in the regression equation. The most authoritative forecasts of social and economic indicators are forecasts of socio-economic development, made by the Ministry of Economic Development of the Russian Federation.

The research focused on the following indicators: installed electric capacity of power plants (N, 10^6 kW), electricity consumption (W, 10^9 kW·h), heat consumption (Q, 10^6 Gcal), population (P, 10^6 people), investments in electric power industry (including district heating) (I, 10^9 RUB), gross domestic product (GDP) (GDP, 10^9 RUB), time interval (T, year). The considered indicators are represented by time series from 1990 to 2014 [11, 12].

An initial analysis of the system of the considered indicators is to identify strong relationship between the indicators in the sample. Correlation analysis allows us to determine if it is necessary to include the indicators in the multiple regression model. The correlation coefficient is a mathematical measure of correlation between two values. We use a linear correlation coefficient (Pearson correlation coefficient) [13, 14], because the indicators are represented by absolute values. Table 1 demonstrates linear correlation coefficients of selected indicators.

Table 1. Matrix of correlation coefficients of the considered indicators from 1990 to 2014.

Indicator	T	N	W	Q	P	I	GDP
T	1						
N	0,8559	1					
W	0,4079	0,6007	1				
Q	-0,928	-0,6522	-0,0667	1			
P	-0,858	-0,5382	-0,3786	0,8358	1		
I	0,8818	0,8946	0,4088	-0,7454	-0,606	1	
GDP	0,987	0,8698	0,4524	-0,8952	-0,8542	0,8905	1

The object of our research is heat consumption, therefore we can draw conclusions about the existence of relationship between heat consumption for the population, GDP and time (the values of the time period). Along with the data of the study, the autocorrelation coefficients of the considered indicators are calculated. Various methods are used to reduce the autocorrelation. The methods aim to exclude the main development trends in the initial data, i.e. linear trends. In this paper, time is introduced in the multiple regression equation as an independent variable [15].

Linear one-parameter regression models of the selected indicators were considered earlier in [1]. Further in the paper, we will consider in more detail the dependence of heat consumption on the selected indicators, by using a multiple linear regression.

2.2 A multiple linear regression model of heat consumption

A linear regression equation in a general form can be written as follows:

$$Y = \beta_0 + \beta_1 X_1 + \beta_2 X_2 + \ldots + \beta_m X_m + \varepsilon \quad (1)$$

where **X** – vector of independent (explanatory) variables, **B** – vector of parameter coefficients (to be determined), ε – random error (deviation), Y – dependent (explained) variable, m – the number of explanatory variables, n – the number of observations.

In order to uniquely solve the problem of finding the regression equation coefficients, the inequality $n \geq m+1$ should be met. To estimate a multiple linear regression, the statistical reliability requires that the number of observations be at least three times greater than the number of indicators to be estimated [13-15]. In our case, in this connection, further calculations are aimed at finding the regression coefficients. Equation (1) in the case of linear multiparameter regression of heat consumption can be written as:

$$y_i = b_0 + b_1 x_{1i} + b_2 x_{2i} + b_3 x_{3i} + b_4 x_{4i} + b_5 x_{5i} + b_6 x_{6i} + e_i \quad (2)$$

where y – values of the explained variable (heat consumption); x – values of explanatory variables ; b – coefficients of the considered linear multiparameter regression; e – values of deviations of sample values of the explainatory variable from the values obtained from the regression equation.

The most common method of estimating the coefficients of the multiparameter linear regression equation is the least-squares method (LSM) [13-16]. The essence of the method consists in minimizing the sum of squares of deviations of the observed values of the dependent variable from its values obtained from the regression equation. Based on our requirement, the standard error should be minimal, which can also be written in the following form:

$$\sum_{i=1}^{n} (y_i - \hat{y}_i)^2 = \sum_{i=1}^{n} (\varepsilon_i)^2 \to \min . \quad (3)$$

Then the equation of multiple regression for description of a change in heat consumption will have the following form:

$$Q_i^{\text{mod}} = 3093{,}67 - 50{,}31 \cdot T_i + 10{,}47 \cdot N_i + 0{,}775 \cdot W_i - 24{,}46 \cdot P_i + 0{,}199 \cdot I_i - 0{,}048 \cdot GDP_i + e_i .$$

A statistical analysis was made to estimate the statistical error, the statistical significance of regression coefficients and the overall quality of the model. The statistical error of the model in the retrospective period was about 1%. The coefficients of the multiple linear regression model are statistically significant. The overall quality of the model is evaluated by determination coefficient R^2. For the considered multiple regression equation, the determination coefficient is 0.994. The closer this ratio to 1, the closer the regression equation describes the behaviour of the dependent variable Y.

The determination coefficient, however, can be quite high in the presence of coinciding trends in the variables under consideration, and as a consequence a high degree of multicollinearity of the variables. The main method to eliminate multicollinearity of the regression model is the method of excluding variables. Further work with the obtained regression model will be aimed at excluding correlated variables.

The studies have indicated that the change in heat consumption is best described by the regression model with two variables (GDP and time), which can be written in the following form:

$$Q_i^{\text{mod}1} = a_0 + a_1 T_i + a_2 GDP_i + e_i^1 . \tag{4}$$

We calculated the values of the regression coefficients a of equation (4) in accordance with the LSM, and this equation has the following form:

$$Q_i^{\text{mod}1} = 1971{,}5436 - 59{,}887 \cdot T_i + 0{,}01297 \cdot GDP_i + \varepsilon_i^1 .$$

A statistical analysis of the heat consumption model (4) was performed. Statistical error of the regression equation is 94.13 or 4.1%. The regression equation coefficients are statistically significant. The determination coefficient of the equation is 0.877. Based on the calculations, we can conclude that the obtained regression equation explains 87.7% of variation in the dependent variable (heat consumption).

2.3 A non-linear regression model of heat consumption

Changes in the heat consumption can also be described by an exponential equation. The variables (regressors) as well as in the previous model, will be represented by time and GDP indicators. For this study, however, we will not use the absolute values of heat consumption and GDP, but the base growth rates of these indicators. The values of the indicators in the final year are chosen as the base value n, due to the fact that logarithm will be made exactly according to the basic growth of heat consumption. The general equation of exponential dependence of heat consumption has the following form:

$$Q^{\text{mod}2} = e^{c_0 + c_1 \cdot T + c_2 \cdot GDP^2} \cdot \varepsilon_i^2 , \tag{5}$$

where $Q_i^2 = \dfrac{Q_i}{Q_n}$ – the basic growth of heat consumption, $i = 1,2,..n$; $GDP_i^2 = \dfrac{GDP_i}{GDP_n}$ – the basic growth of GDP, $i = 1,2,..n$; c_0, c_1, c_2 – the regression equation coefficients.

The LSM is used to calculate the values of the coefficients of equation (5). The equation of heat consumption depending on time and GDP will be as follows:

$$\ln Q_i^{\text{mod}2} = 0{,}2463 - 0{,}0315 \cdot T_i + 0{,}4681 \cdot GDP_i + \ln \varepsilon_i^2 .$$

A statistical analysis was performed for this model of heat consumption. Based on the calculation results we estimated the statistical errors and the statistical significance of the regression equation coefficients. The determination coefficient R^2 was 0.91. These calculations are correct for the logarithm of the basic heat consumption growth. Further, after reduction to exponential function and multiplication by value of heat consumption in base year, the estimated determination coefficient was 0.9071.

3 Case study: heat consumption monitoring and forecasting

The considered regression models quite well describe the change in heat consumption in the retrospective period. These regression models can be used to make a short-term forecast of heat consumption.

In accordance with [17, 18], the forecast of GDP change was adopted and heat consumption levels were calculated using the obtained regression models. Two options of the GDP change forecast were approved, and interval estimates of heat consumption for a prospective period were obtained. The length of the forecast period covered by the regression model depends directly on the retrospective period for which the model is derived. It is assumed that the forecast period should not exceed 1/3 of the retrospective period. According to this circumstance, the heat consumption forecast is made until 2020. Figure 1 presents the results of the heat consumption calculation in accordance with the two options of economic development.

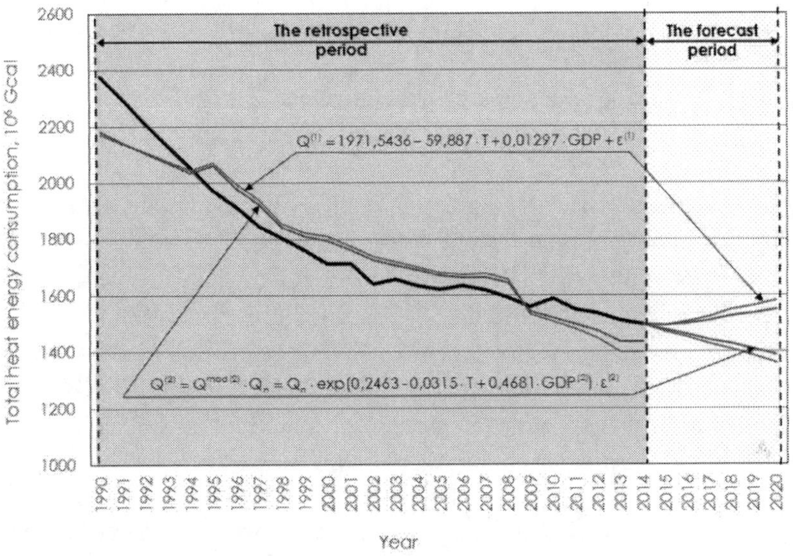

Fig. 1. The use of a multiple linear regression model and non-linear regression model for heat consumption monitoring and forecasting

4 Conclusions

The studies have demonstrated that in accordance with the economic development forecast [17, 18] based on the multiple linear regression model, a decrease in heat consumption by 2020 may be 9.2% in the first option of economic development or it can increase by 5.6% in the second option compared to 2014. According to the exponential model used to predict heat consumption, it can decrease by 2020 to 7.1% in the first option or increase by 3.4% in the second option compared to 2014. The change range of heat consumption, according to the exponential model, has narrower boundaries, which allows us to more accurately determine the possible level of heat consumption. The discrepancy between the projected values of heat consumption at the level of 2020, using linear and nonlinear regression models, is no more than 2% (32×10^6 Gcal). Thus, we can conclude that the obtained regression models can be used for monitoring and medium-term forecasting of heat consumption. In the future, we plan to continue the research in this area. Our plan is to expand the models with the explanatory variables, and describe the structure of heat consumption, using regression models.

The research was performed at Melentiev Energy Systems Institute SB RAS in the framework of scientific projects III.17.4.1 №AAAA-A17-117030310432-9 and III.17.4.3 №AAAA-A17-117030310437-4 of the Fundamental Research Program of SB RAS.

References

1. T.V. Dobrovolskaya. Monitoring of heat consumption levels using regression models. System research in power engineering. Proceedings of young scientists of ISEM SB RAS. **Vol.45** 124-130 (2015)
2. A.M. Mastepanov. *A fuel and energy complex of Russia at the turn of the century: state, problems and prospects*. 793 (2010)
3. A.S. Nekrasov. *An analysis and forecasts of the fuel and energy sector development. Selected works*. 592 (2013)
4. Ming Meng, Dongxiao Niu. Annual electricity consumption analysis and forecasting of China based on few observations methods. Energy conversion an management, **52**, 953-957 (2011)
5. N. Fumo, M.A. Rafe Biswas. Regression analysis for prediction of residential energy consumption. Renewable and sustainable energy reviews. **47**, 332-343 (2015)
6. S. Smeekes, E. Wijler. Macroeconomic forecasting using penalized regression methods. International journal of forecasting. **34**, 408-430 (2018)
7. T. Catalina, V. Iordache, B. Caracaleanu. Multiple regression model for fast prediction of the heating energy demand. Energy and buildings. **57**, 302-312 (2013)
8. Tingting Fang, Risto Lahdelma. Evaluation of a multiple linear regression model and SARIMA model in forecasting heat demand for district heating system. Applied Energy. **179**, 544-552 (2016)
9. G.J. Tsekouras, E.N. Dialynas, N.D. Hatziargyriou, S. Kavatza. A non-linear multivariable regression model for midterm energy forecasting of power systems. Electric power systems research. **77**, 1560-1568 (2007)
10. V.Bianco, O. Manca, S. Nardini. Electricity consumption forecasting in Italy using linear regression models. Energy. **34**, 1413-1421 (2009)
11. Russian statistical yearbook. 2013. 717 (2013)
12. Federal state statistics service (http://www.gks.ru/)
13. E. Ferster, B. Renz. *Methods of correlation and regression analysis*. 302 (1983)
14. S.A. Borodich. *Econometrics: study guide*. 408 (2001)
15. G.A. Ivashchenko, G.S. Kildishev, R.A. Shmoilova. *Statistical study of the main trends of development and interrelation in the ranks of dynamics*. 168 (1985)
16. S.V. Solodusha. *Typical problems of the basic course of econometrics*. 42 (2007)
17. *Energetics of Russia: a view to the future* (Substantiating materials to the Energy strategy of Russia to 2030). 616 (2010)
18. The energy strategy of Russia to 2030 (https://minenergo.gov.ru/node/1026)

An approach to design of an intelligent heating system

Valery Stennikov[1], *Evgeny* Barakhtenko[1] , and *Dmitry* Sokolov[1,*]

[1]Melentiev Energy Systems Institute of Siberian Branch of the Russian Academy of Sciences (ESI SB RAS), Pipeline Energy Systems Department, 130, Lermontov str., Irkutsk, Russia, 664033

Abstract. At present, there is a growing need to solve the design problem of the hydraulic circuit of an intelligent heating system with the required transmission capacity, hydraulic stability and reliability indices. The paper proposes a mathematical statement of the design problem of such a hydraulic circuit and an original methodological approach to solve it by decomposing a general complex problem into several simpler ones (of a smaller dimensionality). The proposed approach makes it possible to tailor the computational procedure to specific features of a heating system to be modeled.

1 Introduction

Heating systems have a great social and economic significance. Nowadays, they have turned into complex spatially distributed pipeline systems for heat supply to the consumer. The complexity of these systems is related to their closed two-line schemes, multi-ring structure, a variety of heat sources and control elements (pumping and throttling stations, regulators, heat points).

A promising technological platform for future heat supply is intelligent heating systems that integrate the sources, networks and consumers at a new technological level into a single automated system [1]. This will provide these systems with the necessary flexibility under their changing operating conditions, enhance heat supply efficiency, reliability, quality; will help reduce heat losses, smooth out uneven heat consumption, etc.

The available possibility for automatic real-time monitoring and control of operating conditions of all participants involved in the process of heat generation, transmission and consumption will enable a quick response to changes in various parameters in the heating system. This will ensure a two-way mutually coordinated interaction between the consumer and the heating system.

In the context of transition to new concepts for the construction of intelligent heating systems, the growing scale and complexity of these systems, and an increase in the number of active network elements (sensors, pressure and flow regulators, etc.), the demand for mathematical models, methods and algorithms intended to design these systems is increasing.

[*] Corresponding author: sokolov_dv@isem.irk.ru

To design a heating system, we solve the problem of ensuring the necessary transmission capacity by determining optimal parameters of the system. A common approach to solving this problem is its reduction to the problem of linear programming [2-4]. The advent of dynamic programming [5] led to a significant development of methods for solving the problems of related to the selection of optimal parameters [6-8]. Some researchers try to use combinatorial methods to solve the problem of optimal design of pipeline systems [9]. In [10], the authors propose the Davidon-Fletcher-Powell method to find a solution. The authors of [11] apply a generalized reduced gradient (GRG) algorithm.

The heuristic algorithms are currently used to develop various approaches to the optimal design of pipeline systems [12-14]. The authors of [15] propose simulated annealing.

In Russia, the method of multi-loop optimization is used to determine the optimal parameters of ring pipeline systems, based on the principle of continuous improvement in solution [16, 8]. This method is further developed in [17-19]. The authors of [8] propose the method of redundant design schemes to select the configuration of pipeline systems.

Design of pipeline systems involves the methods of hydraulic and thermal calculations, which are considered in [20-24]. In [25-26], the focus is made on the methods for an analysis and improvement of energy system reliability.

At present, to solve the design problems of intelligent heating systems we need to develop:
- a single statement of the design problem of a hydraulic circuit of an intelligent heating system with required transmission capacity and indices of hydraulic stability and reliability;
- a single methodological approach to solve the design problem of the hydraulic circuit of an intelligent heating system.

2 Mathematical statement of the problem

The statement of the design problem of the hydraulic circuit of intelligent heating system has the following form: a specified network scheme is a calculated (design) scheme that consists of m nodes and n branches, and is represented by a directed graph $G_{HN} = (J, I)$, where $J = J_C \cup J_S \cup J_B$ – a set of vertices (nodes) that includes a set of consumers J_C, sources J_S and branching points on the scheme J_B; $I = I_P \cup I_S \cup I_C$, where $I_P = I_N \cup I_E$ – a set of branches of linear (pipeline) part of the network that consists of a set of existing I_C and new I_N branches; I_C and I_S – sets of consumer branches and source branches, respectively; $I_{PS} \subset I$ – branches in which the pumping stations are either installed or can be installed. The lengths L_i ($i \in I$) for all the branches are specified.

When solving the problem, we should minimize the total cost function of the heating system, that has the form:

$$F(\mathbf{d},\mathbf{H},\mathbf{x},\mathbf{P}) = \sum_{i\in I_P} F_i^{P}(d_i) + \sum_{i\in I_{PS}} F_i^{PS}(H_i, x_i) + \\ + \sum_{\substack{i\in I \\ \varphi(i)=(j,k)}} F_i^{E}(x_i, d_i, P_j, P_k) + \sum_{j\in J_C} F_i^{C}(P_j) \to \min, \tag{1}$$

where F_i^P – the costs of pipeline construction and operation; F_i^{PS} – costs of pumping station construction and operation; F_i^E and F_j^C – costs of electricity used to pump the transported medium and the costs of its supply to the consumer, respectively ; $\mathbf{d}$ – n-dimensional vector of pipeline diameters; $\mathbf{H}$ – n-dimensional vector of heads at pumping stations; $\mathbf{x}$ – n-dimensional vector of transported medium flow rates in the network branches; $\mathbf{P}$ – m-

dimensional vector of pressure at network nodes; $\varphi(i)$ – function putting each branch $i \in I$ in correspondence with a couple of incident nodes (j,k).

The model of flow distribution in a pipeline network has the following form:

$$\mathbf{Ax} = \mathbf{G}, \tag{2}$$

$$\mathbf{A}^{\mathrm{T}}\mathbf{P} + \mathbf{H} = \mathbf{f}(\mathbf{s},\mathbf{x}), \tag{3}$$

where $\mathbf{A}$ – an $(m \times n)$-matrix of incidence of the calculated scheme; $\mathbf{G}$ – m-dimensional vector of nodal sinks and sources of the transported medium; $\mathbf{f}$ – n-dimensional vector function with elements $f_i(s_i(d_i), x_i) = s_i(d_i)x_i|x_i|^{\beta-1}$, $i \in I$, that reflect the law of pressure drop in the network branch, $s_i(d_i)$ – hydraulic resistance of pipeline.

A system of constraints consists of the following components:

$$P_j^{\min} \le P_j \le P_j^{\max}, \quad j \in J; \tag{4}$$

$$d_i \in D, \quad i \in I_{\mathrm{P}}; \tag{5}$$

$$H_i \in H_{\mathrm{S}}, \quad i \in I_{\mathrm{PS}}; \tag{6}$$

$$v_i^{\min} \le v_i(x_i, d_i) \le v_i^{\max}, \quad i \in I_{\mathrm{P}}; \tag{7}$$

where $P_j^{\min}$ and $P_j^{\max}$ – lower and upper pressure constraints; D – a set of diameters of standard pipelines; $v_i^{\min}$ and $v_i^{\max}$ – lower and upper constraints on the transported medium velocity v_i; H_{S} – a set of heads at pumping stations that can be installed.

The reliability requirements impose the constraints that are determined by standard values of:

- availability factor at all consumer nodes

$$K_j \ge K^{\min}, \quad j \in J_{\mathrm{C}}, \tag{8}$$

- failure-free operation probability index for all consumer nodes

$$R_j \ge R^{\min}, \quad j \in J_{\mathrm{C}}, \tag{9}$$

where $K^{\min}$ and $R^{\min}$ are lower constraints on the availability factor and probability of failure-free operation, respectively.

Constraints on the required level of hydraulic stability for each consumer are

$$\omega_i(x_i, G_j, G_k) \ge \omega_i^{\min}, \quad i \in I_{\mathrm{C}}, \quad \varphi(i) = (j,k), \tag{10}$$

where ω_i – function of hydraulic stability of the consumer, $\omega_i^{\min}$ – required value of a coefficient of hydraulic stability of the consumer.

By solving the problem of function minimization (1), subject to (2)-(10), we should determine the parameters of the intelligent heating system.

3 A proposed approach to solve the problem

We propose an original methodological approach to solve the stated design problem of the hydraulic circuit of the intelligent heating system. The approach is based on the concept of decomposition of a general complex problem into a set of less complex problems. By decomposing the general problem we obtain the following conceptual problems to solve:

- select the network configuration;
- determine optimal parameters of the hydraulic circuit;
- estimate and ensure the necessary level of system reliability;
- carry out hydraulic and thermal calculations.

The problem of selection of the network configuration lies in determining the best graph of the system. It is solved by the method of redundant design scheme [9]. The determination of optimal hydraulic circuit parameters means the search for a solution ensuring the best transmission capacity of the network by replacing pipelines and installing (retrofitting) pumping stations. This problem is solved by the methods of dynamic programming and multi-loop optimization [17, 18]. The problems of hydraulic and thermal calculations are reduced to solving systems of linear and nonlinear equations. These problems are solved by the effective methods developed at Melentiev Energy Systems Institute of SB RAS [8, 23, 24].

The proposed methodological approach suggests the following steps to solve the design problem of hydraulic circuit of intelligent heating system (Fig.1):

Step 1. Build a redundant design scheme of a heating system.

Step 2. In the case that there is no need to solve the problem of network configuration selection, go to step 4.

Step 3. Solve the problem of network configuration selection.

Step 4. Determine optimal parameters of the hydraulic circuit.

Step 5. Calculate and assess reliability indices.

Step 6. In the case that the required reliability level is provided, go to step 8.

Step 7. Solve the problem dealing with addition of backup ties to provide the required reliability level. If these ties are necessary – go to step 1, if not – go to step 3.

Step 8. Calculate and analyze hydraulic and thermal conditions.

Step 9. Based on the analysis of hydraulic and thermal conditions make a decision on the correction of the parameters. If the correction is necessary, go to step 4.

Step 10. Calculate and analyze the cost-effectiveness indices.

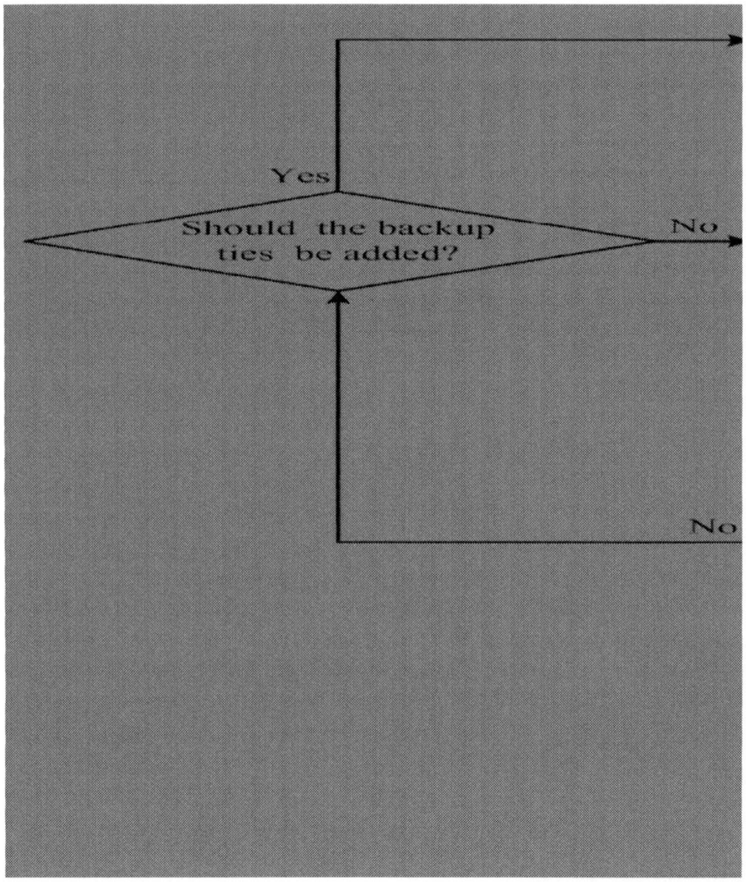

Fig. 1. A proposed algorithm.

The developed methodology was applied to investigate the possibility of construction of an intelligent heating system in the district Novo-Lenino of the city of Irkutsk, which is a typical urban district with prevailing residential housing. The solutions obtained from the calculation are intended to develop this system and transform it into an intelligent system to provide better comfort, quality, and reliability of heat supply to consumers in this district.

4 Conclusion

We propose a mathematical statement of the design problem of hydraulic circuit of intelligent heating system with the required transmission capacity and necessary indices of hydraulic stability and reliability. An original methodological approach is proposed to solve the formulated problem. This approach suggests decomposition of a general complex problem into several less complex problems (of smaller dimensionality and complexity).

The proposed methodological approach makes it possible to:

- Fully consider the specific characteristics of equipment applied, ways of its installation and operation;
- Use complex nonlinear mathematical models;
- Adjust computational procedure to specific features of a system to be modeled.

The developed methodology can be used to make recommendations to improve and transform complex heating systems into intelligent systems, enhance their efficiency and quality of heat supply to consumers.

The research was supported by Russian Science Foundation (grant No. 17-19-01209).

References

1. H. Lund, S. Werner, R. Wiltshire, S. Svendsen, J. E. Thorsen, F. Hvelplund, Energy, **68**, 1-11 (2014)
2. E. Alperovits, U. Shamir, Water Resour. Res. **13**, 885-900 (1977)
3. I. Goulter, J. Water. Res. Plan. Manage. Div. Am. Soc. Civ. Eng., **118**, 238-248 (1992)
4. G. Eiger, U. Shamir, A. Ben-Tal, Water Resour. Res. **30**, 2637-2646 (1994)
5. R. E. Bellman, *Dynamic programming* (1957)
6. J. R. Appleyard, Build. Serv. Eng. **45**, 191-204 (1978)
7. L. Garbai, L. Mobnar, Colloq. math, societatic Janos Bolyai. **12**, 373-390 (1974)
8. A. Merenkov, V. Khasilev, *The Theory of Hydraulic Circuits* (1985)
9. J. Gessler, *Proc. of the Ninth International symposium on Urban Hydrology, Hydraulics and Sediment Control*, (1982)
10. A. Chiplunkar, S. Mehndiratta, P. Khanna, J. Env. Eng. Div. Am. Soc. Civ. Eng., **112**, 264-279 (1986)
11. K. Lansey, L. Mays, J. Hydraul. Div. Am. Soc. Civ. Eng. **115**, 1401-1418 (1989)
12. K. McClymont, E. Keedwell, D. Savic, Environ Modell Softw, **69**, 414-424 (2015)
13. D. Savic, G. Walters, J. Water Resour. Plann. Manag. **123**, 67-77 (1997)
14. J. W. Nicklow, P. M. Reed, D. Savic, T. Dessalegne, L. Harrell, A. Chan-Hilton, M. Karamouz, B. Minsker, A. Ostfeld, A. Singh, E. Zechman, J. Water Res. Plan. Man. **136 (4)**, 412-432 (2010)
15. M. Cunha, J. Sousa, J. Water Res. Plan. Manage. Div. Soc. Civ. Eng., **125**, 215-221 (1999)

16. S. Sumarokov, Ekonomika i matematicheskie metody, **12**, 1016-1018 (1976)
17. V. Stennikov, E. Barakhtenko, D. Sokolov, T. Oshchepkova, *Sustaining Power Resources through Energy Optimization and Engineering Premier Reference Source*, 76-101 (2016)
18. V. Stennikov, E. Barakhtenko, D. Sokolov, Therm Eng, **64**, 518-525 (2017)
19. D. Sokolov, V. Stennikov, T. Oshchepkova, E. Barakhtenko, Therm Eng, **59**, 337-343 (2012)
20. D. W. Martin, G. Peters, IWES J., 1963
21. U. Shamir, C. D. D. Howard, J. Hydraul. Div. ASCE, **94** (1968)
22. E. Todini, S. Pilati, System Analysis and Simulation, 1-20 (1987)
23. N. Novitskii, Therm Eng, **60**, 1051–1060 (2013)
24. V. Tokarev, Z. Shalaginova, Therm Eng, **63**, 68-77 (2016)
25. M. Haghifam, M. Manbachi, Int J Elec Power, **33**, 385-393 (2011)
26. I. Postnikov, V. Stennikov, E. Mednikova, A. Penkovskii, Energy Procedia, **105**, 3083-3088 (2017)

Problems of organization of operating conditions of large heat supply systems and their solutions based on multilevel modeling†

Zoya I. Shalaginiva[1], and *Vyacheslav* V. Tokarev[1]

[1]Melentiev Energy Systems Institute of Siberian Branch of the Russian Academy of Sciences (ESI SB RAS), Pipeline Energy Systems Department, 130, Lermontov str., Irkutsk, Russia, 664033

Abstract. The paper unveils relevance of applying the methodology of multi-level modeling of thermal-hydraulic conditions of heat supply systems (HSS) and the software which are realized in the information-computing system "ANGARA-TS", which make possible development of operating conditions and adjustment measures in HSS of any structure and complexity including non-standard circuit designs. Application of the multi-level modeling of HSS that is based on integration of mathematical models, methods and information technologies allows one to link the main principle of mathematical modeling that contributes to rational correspondence of the degree of mathematical model detail with the goals of its use, and modeling for different purposes. Information-computing system is implemented in a variety of real HSS of large cities in the organization of regimes and development of adjustment activities. The paper describes the experience of using new methods for development of operating conditions of large HSS in real towns.

1. Introduction

The present day HSS of cities are complex spatial high-dimensional objects which are developing in time and combine a set of heterogeneous elements. Frequently several heat sources supply heat to looped heat networks with the installed pumping stations, central heat points, pressure controllers and other control devices. All these factors complicate problems of operation, and control of HSS, planning and development their conditions.

Over the past decade the volumes and structure of heat consumption as well as the structure of systems themselves have changed considerably. The emerging new trends of semi-voluntary (without adequate validation) connection of new townhouse villages with low heat loads and direct connection of heating systems substantially complicate development of thermal-hydraulic conditions of HSS. This is explained by the fact that the excess heads at the points of connection to these houses cannot be reduced technically even with the minimum diameters of feed pipelines and installation of throttlers at the points of connection with minimum diameters. As a consequence, the higher (in comparison with the

† The study was carried out within the framework of the III.17.4.3 project of the Basic Research Program of the SB RAS (AAAA-A17-117030310437-4) with the financial support of the Russian Foundation for Basic Research and the Government of the Irkutsk Region within the framework of the scientific project no. 17-48-380021.

required) heat carrier flows circulate through such consumers, which results in overheating of the considered houses, increase of water temperature in the return main pipeline, high heat losses (due to discharge of the unspent heat carrierз into the return main pipeline) and in essential increase of circulating flow rates in the whole system. Consumers connected close to the heat source influence operating conditions most significantly.

General ageing of equipment installed in HSS makes it impossible to maintain the target temperature charts. At the same time transition to a low temperature chart is impossible because of insufficient throughput of heat networks which were originally designed for a higher temperature chart. Increase of head losses in local heat consumption systems that is caused by their clogging because of absence of planned cleaning and overhauls of the internal distribution pipelines leads to the fact that the hydraulic jet pumps installed at the connection points cannot ensure a required mixing rate even with sufficient available heads.

In view of the permanently rising prices of thermal energy, consumers actively install heat meters and autonomous control devices. As a result, the thermal-hydraulic conditions of the system change and the processes of system operation and control of conditions become more complicated. The traditional techniques, however, are not adapted to development of operating conditions and adjustment measures in such HSS, where some consumers have manual connection nodes, other consumers are equipped with autonomous control devices. The techniques for fully automated systems are unavailable as well.

Changes in administrative and economic management brought about direct interests of operating enterprises in economical consumption of heat carriers and heat, and its maximum efficient use. This requires the adjustment measures to be carried out before each heating season, as well as development of operating conditions of HSS to guarantee their reliability and quality. Conventionally, specialized organizations were engaged in development of the adjustment measures, tests of networks and calculations of different conditions. However, recently despite the reference documents, which include the need for annual adjustment of networks, such works have been performed irregularly. Often the issue about their execution and financing arises after emergencies and considerable material and losses.

2. Purposes and tasks of work

The currently formed operation environment of HSS motivates topicality of applying the methodology of multi-level modeling developed at MESI SB RAS, special methods and techniques to calculate normal operating conditions in terms of the above factors [1-7].

The purpose of the work was: approbation of a multilevel approach to modeling the conditions of large heat supply systems hierarchical structures on real HSS, the definition of bottlenecks and the specifics of the organization of their conditions. The novelty of the work is to generalize the problems of the organization of HSS conditions and to modify the methods and algorithms of calculation in order to automate their solution. To overcome the difficulties of regimes organization, it was necessary to develop both a methodological base and software tools for adjustment of thermal-hydraulic conditions and calculation of throttling devices.

3. Means and approaches to the development of HSS conditions

The Laboratory of Pipeline and Hydraulic Systems at MESI SB RAS has been engaged for some years in theoretical and applied studies in the area of modeling, analysis and optimization of HSS [8-19]. The information-computing system (ICS) "ANGARA-TS" [20-21] is a latest design that is an integration of information-computing environment [4, 7] and program-computing systems [1, 5] to develop operating conditions of heat networks (normal, repair, post-emergency). The multi-level modeling of HSS, which is the basis of the

ICS, enables to overcome the problems of: fragmented initial information, provision with methods, models, and algorithms to solve different tasks arising at the stage of planning and development of conditions and adjustment measures; high dimensionality of current HSS; efforts to validate quantitatively and coordinate solutions on HSS control. Moreover, it enables to link the basic principle of mathematical modeling, which ensures rational correspondence of the degree of the mathematical model detail with the goals of its application, and perform modeling for different purposes. This new information-computational technology satisfies main requirements for creation of the automated control systems of district heating objects "Heat networks" [22]. The technology is a qualitatively new generation of software [17], which integrates information functions and a practically full set of computational modules, needed for solving the problems of adjustment and operation of HSS, planning, development and analysis of operation conditions of heat networks. The list of problems solved using the ICS is presented in [17].

4. Specific features of problems on development of HSS conditions

Consider the issues of development of operation conditions of HSS by the example of real towns which were resolved by the efforts of MESI SB RAS using the ICS "ANGARA-TS". HSS of modern cities are characterized by highly heterogeneous connection schemes of consumers to heat networks: automated and manual connection nodes; dependent (direct, through hydraulic jet pumps, with a mixing pump) and independent connection of heating systems; open and closed water extraction for hot water supply; different connection schemes of heaters for hot water supply (single-stage – parallel or series; double-stage – series or mixed); water extractions for hot water supply from heating networks; availability (absence) of heat metering units; great scatter of load values and geodesic marks of consumers; multi-stage control, etc. Hence, development of conditions was connected with the following main difficulties.

1. Connection of consumers to heat networks through the hydraulic jet pumps of heating systems requires high available heads at the connection nodes. At the same time direct connection of private houses with a low heat load to these heat networks makes it impossible to decrease high excess heads at their connection nodes that are conditioned by the hydraulic jet pumps. Therefore, the multi-stage decrease of heads in the network and at consumers called for development of nonstandard solutions (Fig.1).

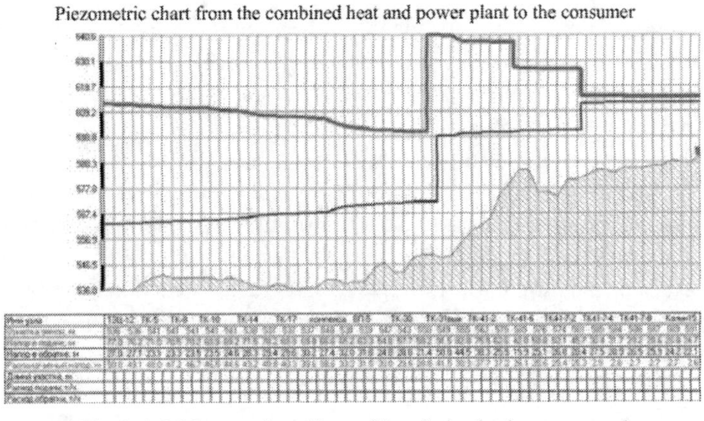

Fig.1. Multistage throttling of heads in the heat network.

2. Low velocities of a heat carrier in distribution networks which connect houses with a low heat load led to large heat carrier cooling down and impossibility of serving the required heat load to consumers with the existing hydraulic condition. In the general case the heat losses in heat networks can be compensated by: increasing the flow rates with temperature preservation and making the corrections in the temperature chart. Transition to a higher temperature chart should be justified economically and is not always possible because of equipment state. Therefore, if the throughput of heat networks is capable of cover-

ing heat losses by increasing the flow rate, the heat carrier temperature was no corrected. Otherwise, the temperature chart was corrected [18].

3. In some towns insufficient throughput of heat networks to supply heat to consumers based on the temperature chart 95/70°C necessitated transition to the higher temperature chart 150/70°C that was initially accepted as a design chart for heat networks. On the other hand, the wear of heat source equipment did not allow the support of this curve. In addition, transition to a higher temperature curve required the mixing points to decrease the temperature of water coming to local heating systems which were dismantled because of their work based on the lower curve for years. The formed situation demanded justification and development of temperature curves. The formed situation demanded justification and development of the temperature curve with cutting during the period of low temperatures (Fig.2).

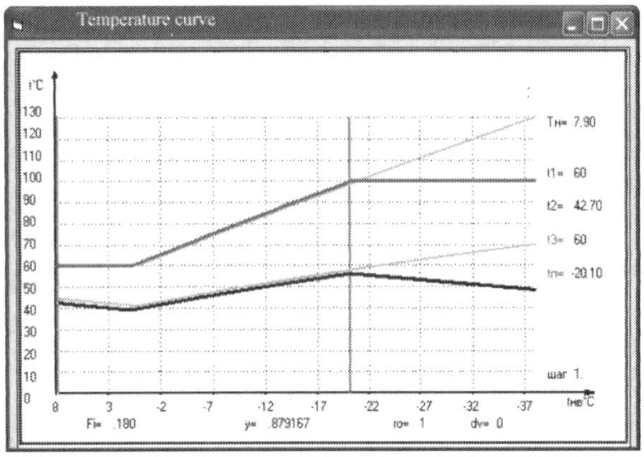

Fig.2. Temperature curve of heat supply from CP-12 in the Cheremkhovo town.

4. Availability of group mixing nodes did not always provide a required mixing ratio because of high head losses in distribution networks.

5. Availability of water withdrawal for hot water supply from heating networks (in the closed HSS with four laid pipes) at some consumers necessitated refinement of the technique for adjustment calculation of thermal-hydraulic condition subject to unauthorized water withdrawals.

6. Rugged topography, which narrows an admissible region of decisions made, demanded multivariant calculations of thermal-hydraulic conditions using multistage throttling on the network and at consumers (see Fig.1).

7. Absence of funds for proper reconstruction of heat networks and upgrading of central heat points.

5. Development of methodological base for solving the problem of operating conditions calculation for HSS

The indicated difficulties in development of conditions could be overcome by evolution of both the methodology and software for adjustment calculations of thermal-hydraulic conditions and calculation of throttlers. Within the framework of the ICS "ANGARA-TS" the following techniques were developed and applied.

1. The technique of the single-and multi-level adjustment calculation of thermal-hydraulic condition of HSS subject to heterogeneity of connection schemes of consumers to heat networks [23–25]. It makes it possible to supply a required amount of heat to consumers, provided the constraints on heat carrier parameters at all system nodes are observed (Fig. 3). The technique is based on the calculation of corrections to heat carrier flow rates that are differentiated for each consumer and required to compensate for heat losses in heat networks. Besides, it is used to determine parameters of the throttlers simultaneously for all consumers and in the heat network. The technique enables to automatically obtain the coordinated operation conditions of heat networks of different levels with multi-stage heat supply, to significantly improve development of conditions and to reduce the time of heat network adjustment by far.

2. The technique for determination of master consumers to calculate heat carrier parameters at the control nodes: make-up water pressure, available heads of heat sources, cen-

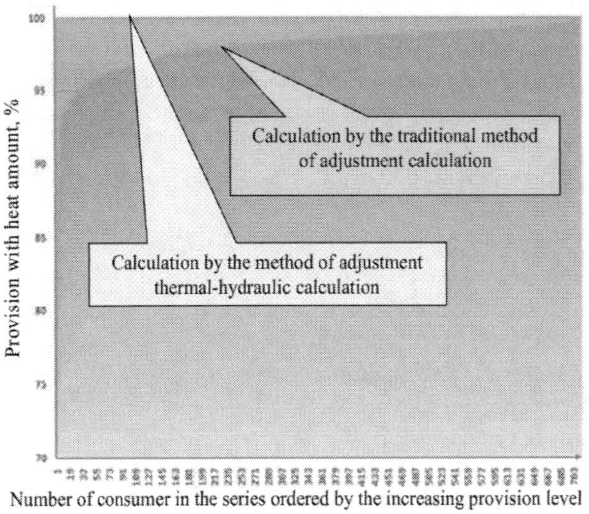

Fig. 3. Graph of consumer provision with heat load.

tral heat points and pumping stations. The technique is based on the analysis of condition feasibility [26-27] and allows the automated determination of the minimum necessary values of parameters of control elements.

3. The technique of adjustment calculation of thermal-hydraulic conditions of hot water supply networks which allows one to determine individual corrections of circulating flow rates of consumers to support the required temperatures at the takeoff nodes. The technique assures required temperatures and water use for hot water supply in accordance with the Construction Rules and Regulations and contractual loads, valid circulating flow rates and substantially decreases water discharges by population [25].

4. The method of multi-stage throttling in heat networks and at consumers, which is topical for highly rugged relief (Fig.1).

5. The technique of searching for a rational flow distribution by sectioning of heat networks with a multi-loop structure. The technique is based on the goal-oriented enumeration of the variants of cleavage close to the nodes of flow junction with further correction of the solution. The solution is corrected in two stages based on the criterion of the total excess hydraulic power [28].

6. The effectiveness of the proposed approaches to the operating conditions calculation

The methods and algorithms applied in the ICS were extensively tested in calculations of HSS in the following towns and cities: Irkutsk, Angarsk, Bratsk, Shelekhov, Ust-Kut, Cheremkhovo, Kiev, Moscow, St. Petersburg, Ulan-Bator, Tyumen, Tambov, Petropavlovsk-Kamchatsky, etc. This became possible owing to the long-standing close cooperation with some design and operation institutions, research institutes and universities.

Development of conditions and adjustment measures on the basis of the suggested technology of calculations improved quality of heat supply to towns and simultaneously to identify and realize the great potential of energy saving due to reduction of: circulating flow rates; make-up water flow rate and water discharges by population; heat losses; electricity consumption for heat carrier pumping and chemical water treatment; switched some distribution networks to operation by a lower temperature curve, which became possible owing to substantial decrease of circulating flow rates after elaboration of adjustment measures.

The annual economic effect of the development of conditions and adjustment measures in the cities of the Irkutsk region amounted to, million rubles: Irkutsk - 274; Angarsk -125; Bratsk -43; Cheremkhovo - 9.

7. Conclusion

1. The work identifies and summarizes the problems of the development of operating conditions HSS.

2. A set of methods and the technology of calculations applied in the ICS are suggested. They make it possible to elaborate adjustment measures in HSS of any structure and complexity, including those with nonstandard schemes. This made it possible to significantly reduce setup time and achieve significant improvements in the development of operating conditions with limited funds for network re-routing and system upgrading.

3. Application of the multi-level modeling allowed the calculations of conditions of large HSS with intermediate control steps.

4. The developed techniques of adjustment calculation substantially decreased an adjustment period and significantly improved arrangement of conditions subject to limited funds for removal of existing networks and modernization of systems.

5. The suggested methodology was applied to development of conditions and adjustment measures in the heat supply systems of real cities. Implementation of the planned measures improved quality of heat supply to consumers without capital investment in networks reconstruction and essentially decreased circulating flow rates.

References

1. N.N. Novitsky, V.V. Tokarev, Z.I. Shalagonova et al., *Research and development of the SB RAS in the field of energy-efficient technologies*, 48 (Novosibirsk: Nauka, 2009).
2. N.N. Novitsky, A.V. Alekseev, O.A., Grebneva et al., *Information and mathematical technologies in science and management*, 110 (Novosibirsk: Nauka, 2007).
3. A.V. Alekseev et al., *Pipeline systems in energy sector*, 211 (Novosibirsk: Nauka, 2008).
4. A.V. Alekseev et al., *Pipeline systems in energy sector*, 221 (Novosibirsk: Nauka, 2007).
5. N.N. Novitsky, V.V. Tokarev, Z.I. Shalagonova et al., Information and mathematical technologies in science, engineering and education, **X (II)**, 285 (Irkutsk, ESI SB RAS, 2005).
6. Z.I. Shalagonova, N.N. Novitsky, V.V. Tokarev, O.A., Grebneva *Energy of Russia in the XXI century. Innovation development and management*, 389 (Irkutsk: ESI SB RAS, 2015).
7. Alekseev, N.N. Novitsky, Z.I. Shalaginiva, Informstion and mathematical technologies in scientific research, **XI**, 7 (Irkutsk ESI SB RAS, 2006).
8. N.N. Novitsky et al., *Pipeline systems in energy sector,* (Novosibirsk: Nauka, 2015).
9. N.N. Novitsky, Thermal Engineering, **56 (12)**, 1010 (2009).
10. V.G.Sidler Improvement of control and automation of heating systems, **VIII**, 37 (Kharkov, 1988).
11. Z.I.Shalaginova, Doctoral thesis in engineering, (Irkutsk: ESI SB RAS, 1995).
12. Z.I. Shalaginova, Thermal Engineering. **56 (12)**, 1016 (2009).
13. Z.I. Shalaginova, Human-Environment-Universe, 182 (Irkutsk: IrSTU, 1997).
14. N.N. Novitsky, V.V. Tokarev, Z.I. Shalaginova, Information systems and technologies in energy sector and housing and public utilities, 50 (Ukraine, Kharkov: KHNURE, 2011).
15. Z.I. Shalaginova, Mathematical models and methods for analysis and optimal synthesis of developing pipeline and hydraulic systems, **XII**, 315 (Irkutsk: ESI SB RAS, 2010).
16. Z.I. Shalaginova, Thermal Engineering, **63 (3)**, 222 (2016).
17. N.N. Novitsky, Z.I. Shalaginova, V.V. Tokarev, Izvestiya RAN: Energetika. **1**, 12 (2018).
18. Z.I Shalaginova, Thermal Engineering, **51 (7)**, 554 (2004).
19. Z.I Shalaginova, *Pipeline systems in energy sector*, 372 (Novosibirsk: Nauka, 2004).
20. N.N. Novitsky, A.V. Alekseev et al., The Proceedings of the International Scientific Conference on Power industry and market economy, 323 (Mongolia, Ulaanbaatar, 2005).
21. N.N. Novitsky et al., Energy. (in press) https://doi.org/10.1016/j.energy.2018.02.070
22. RD 34.35.127-96 (Moscow, 1996).
23. V.V. Tokarev, Z.I. Shalaginiva, Mathematical models and methods for analysis and optimal synthesis of developing pipeline systems, **XII,** 300 (Irkutsk: ESI SB RAS, 2010).
24. V.V. Tokarev, Z.I Shalaginova, Thermal Engineering, **63 (1),** 68 (2016).
25. Z.I. Shalaginiva, V.V. Tokarev, O.A. Grebneva, Proceedings of IrSTU, **3 (98)**, 165 (2015).
26. A.V. Lutsenko, Systems studies in energy, **46**, 28 (Irkutsk: ESI SB RAS, 2016).
27. N.N.Novitsky, A.V. Lutsenko, Information and mathematical technologies in science and management, **XX**, 138 (Irkutsk: ESI SB RAS, 2015).
28. V. V. Tokarev, Thermal Engineering. **65 (6)**, 400 (2018).

Information and computing spaces integration for operating enterprises of pipeline systems using "ANGARA" computer technology*

Aleksandr V. Alekseev[1,†], and *Nikolay* N. Novitsky[1]

[1] Melentiev Energy Systems Institute of Siberian Branch of the Russian Academy of Sciences (ESI SB RAS), Pipeline Energy Systems Department, 130, Lermontov str., Irkutsk, Russia, 664033

Abstract. The article describes the informatization level of enterprises operating pipeline networks. Existing problems are presented. These problems are related to the fragmentation of information systems and the inability to provide information support for decision-making on the management of the development and operation of pipeline systems. An overview of the most high-usage used software for automate the activities of enterprises is given. The need to create a common information space of an enterprise and the opportunities it provides is shown. A brief description of the information-computer environment "ANGARA" and the technology of organizing a common information space on its basis is given. The experience of practical application of the information-computer environment "ANGARA" for creating the common information space of the Municipal Unitary Enterprise "Vodocanal" in Irkutsk is cited. On the diagram of the organizational structure of the enterprise the workplaces with access to a common information is showed.

Introduction

Most of domestic operating enterprises and organizations have a "focal" nature of informatization. In separate subdivision and services of the unified organization, independent, disparate databases (DBs) and information systems are created that are related to the same object (pipeline system) of operation, control or design. As a result, there is a duplication of information in different databases, its inconsistency, unproductive efforts for its support and coordination both in different databases and in different departments of the organization. At the same time many important data for decision-making is absence or doubtful. The problem solving of "focal" automation lies in the plane of creating of the common information space of the enterprise (CISE), which provides the opportunity for rapid access for all services and specialists to any information available at the enterprise. Thus, one of the primary goals in the field of informatization of operational enterprises should be the origination of the CISE. Taking into account the emerging trends towards intellectualization and integration of pipeline networks, the development of the CISE is a necessary condition for the application of new management concepts.

* The research was carried out within the project III.17.4.3 of the Fundamental research program of SB RAS (AAAA-A17-117030310437-4) with finance support of RFBR and the Government of Irkutsk Region in the framework of research project № 17-48-380021

† Corresponding author: asanya@mail.ru

The main objects of the common information space [1] are: information resources - databases, arrays of documents, all types of archives of schemes, project documents, etc. containing data recorded on appropriate media.

The CISE is designed to eliminate inconsistency, duplication of information, the need for parallel data support in various information systems (IS) by automatically synchronizing databases of various IS. This will reduce the organizational barriers between the various structural divisions of the enterprise (or between different enterprises) through the rapid exchange of electronic data, as well as reduce the time and labor costs for document circulation. Thus, it is possible to increase the controllability of the enterprise and the quality of the management decisions taken to develop and operate the pipeline systems.

The CISE should provide the following opportunities: 1) completeness of information about the pipeline system by integrating various information resources (cartographic, schematic, parametric, passports, operational, contractual, etc.); 2) accuracy of data - by eliminating its duplication or inconsistency, either through the principle of a "single data entry point" or by differentiating access rights; 3) availability - by providing on-line access to both enterprise specialists and external users of information in local or global network; 4) end-to-end information support for solving tasks of different territorial and temporal levels of management (automation, reconstruction, development, design, dispatching control, operation, e.t.c.) on a common information basis; 5) the flexibility of the development and application of the CISE ¬ by ensuring the possibility of integrating of new IP, updating or replacing existing ones. [2]

There are several software technologies for the CISE organization [3]:

- within a single software platform (for example, Microsoft products are integrated with each other via OLE / ActiveX / API tools);
- synchronization of databases of various IS;
- centralization of information through a single system, in which a complete set of all data is presented;
- organization of access to the information that is stored in distributed databases of different IS by means of their integration with each other (COM, CORBA, SOA, etc.).

Depending on the IS installed in the enterprise, the most appropriate technology or their combination is selected when organizing the CISE.

Brief description of the current state of informatization of water utilities and existing software

Enterprises operating pipeline engineering networks of cities usually have a number of IS such as billing systems (settlement systems with subscribers), financial and personnel management and accounting systems, geographic information system, CAD systems, telemetry systems (transmission of measured process parameters), passport systems, electronic design model, dispatching logs, etc. Each of these systems has its own data format. This so-called "focal" or "patchwork" automation leads to duplication and inconsistency of data, lack of access to the data for decision-makers on the control of pipeline systems.

Attempts to overcome such disunity are being made, and the need for the CISE organization is actively discussed, both by IT specialists [4-6] and by specialists in operational enterprises [7-10]. However, at the same time, the problem of CISE creating is often identified with the automation of document circulation in order to optimize the production and economic activities of enterprises (business processes). To date, a number of general purpose ISs for automating such processes (ECM - Enterprise Content Management) such as Documentum, Directum, Efrat, 1C: Document Management, Curator, etc. have been developed and implemented.

In many water utilities universal software program are used to automate the activities of the entire enterprise. For example, Municipal Unitary Enterprise (MUE) "Vodokanal of St. Petersburg" and MUE "MosVodokanal" use a unified information system for the enterprise based on Oracle E-Business Suite (OEBS) technology [11] - a full-featured complex of integrated business applications of ERP II-class, other companies use SAP, 1C, Galaxy, Sail, Curator and so on. A brief description of them is given below.

OEBS is designed to automate the main activities of enterprises: the effectiveness of financial management, material flows, relationships with customers and suppliers, personnel, logistics, marketing, sales, maintenance and repair, and others. OEBS provides: transparency of business processes related to purchases and inventories; streamlining the planning of repairs and providing parts with spare parts and materials; improving the internal control of key business processes; improving the quality and reducing the time required for reporting. The system allows solving some technological problems; however it is not focused on the complex decision of technological control problems, has a high cost, both the software product and its implementation.

"1C Enterprise" [12], with various additional configurations ("1C: Enterprise 8. Vodokanal Management" or "1C: Enterprise 8. Technical Settlement Center of the Heating Network") provides management capabilities: production, finance, sales, purchases, storage (stocks), fixed assets and repairs, personnel and payroll, relations with customers and suppliers, monitoring and evaluation of the work of managers, monitoring and analysis of the activities of the enterprise, etc. Accounting for technological features of pipeline systems (PLS) are: equipment certification, calculation of losses in trunk and district networks, formation of water balance, automatic calculation of water supply and sewage reception and formation of payment documents in accordance with accepted calculation methods. The system "1C Enterprise" weakly takes into account the technological features of PLSs and does not have the possibility of their graphic display.

"Galaxy ERP" is a full-scale information system developed specifically for large and medium-sized enterprises in Russia and ex-USSR countries. It allows automating the activities of both the individual enterprise and the holding as a whole, being the basis of the CISE [13]. The main functional subsystems of the "Galaxy ERP" are: production management and planning, production asset management, financial planning, managerial accounting, tax and warehouse accounting, logistics management, personnel policy, security and comprehensive data analysis. In comparison with the "1C" "Galaxy ERP" system is oriented to larger enterprises and has fewer configuration options, and all improvements must be ordered from the developer, which is not always acceptable due to significant financial and time costs.

All mentioned and others ISs are classified as ERP and MRP categories and do not meet the requirements for the CISE. They are primarily aimed on business processes optimizing, automating workflow, accounting tasks, financial activities of an enterprise, statistical analysis and paying insufficient attention to technological processes of water supply and distribution, having only a few modules for solving specific technological problems.

Formation of the CISE based on the information-computer environment "ANGARA"

In the laboratory of pipeline and hydraulic systems of the Energy Systems Institute (ESI SB RAS, Irkutsk, Russia) the universal information and computing environment (ICE) "ANGARA" [14-19] was developed. It designed to automate the creation, support and use of electronic models of PLSs of various types and complexity in solving the applied problems of their research, design and operation within a common graphical user interface. ICE provides the following main features: 1) setting up the user's environment (configuring the structure, filling and integration of databases; 2) setting up the computing environment (integration of software and computational units and complexes); 3) application of this environment within the framework of a common graphical user interface for computer modeling of PLSs [19].

The result of the ICE configuration is the finite information and computing complexes for solving the PLS simulation problems. The obtained complexes frees the user from the need of knowledge of programming, calculation methods and mathematical modeling, allows solving substantive engineering tasks directly on the calculation schemes, using the usual concepts and objects. Thus, ICE provides the ability to create and work with a computer graphics model of real PLS of any type and purpose.

ICE "ANGARA" can potentially be used as a core, both when creating the CISE, and for its use in computer networks, while ensuring a relatively low cost of work for the integration of disparate subsystems of the automated process control system with automated industrial management system. In particular, due to the openness and extensibility of the databases, it is possible to integrate the database of ICE "ANGARA" with such IS as geographic information system, telemetry, commercial accounting, contractual conditions, etc. Such integration can be realized in the following ways: 1) creation of correspondence of records in tables of ICE "ANGARA" and adjacent ISs ("one to one" or "many to many"); 2) ICE inquires the relevant applications or their databases and writes the received information to it`s own database or displays the necessary data on the screen according to the user request; 3) third-party applications transmit the data (when it appear or change) or request them from the ICE "ANGARA" database. Such automation can be carried out without the participation of ICE "ANGARA" developers.

Experience of the CISE creation in MUE "Vodokanal" of Irkutsk city.

When creating an automated workstation of dispatcher in MUE "Vodokanal" in Irkutsk [18,20], a prototype of the CISE for three types of PLS (water supply, sanitation and district heating) was made based on the ICE "ANGARA" (fig.1). IS "1C Enterprise 8. Management of the Vodokanal" and IS "Curator" had been installed at the company. As a result of the implementation the following capabilities were provided: 1) support (display, editing, activation, modeling, etc.) of different types of pipeline systems on a common cartographic basis; 2) updating of the geographic information (urban plan) from the monthly updated city GIS system; 3) updating data on contractual and actual loads of consumers, through integration with the ERP consumer work system "1C Enterprise"; access to other data on consumers (contracts, management companies, arrears, etc.); 4) integration with dispatching electronic journal (log) allows to maintain the current state of the valves in design schema; 5) integration with SCADA system allows using actual measured network parameters in hydraulic calculations; 6) obtaining data on real water consumption directly from accounting nodes by integration with billing systems; 7) storage of project documentation in the ICE "ANGARA" database; 8) carrying out various calculations using all the data available in the CISE (clauses 1-7), not only the data from the ICE "ANGARA" database [2].

The main difficulty in developing the CISE was to create a correspondence between the records in different IS. There is a discrepancy between the number of records in different databases, and the addresses and names of objects, which forces us to develop intermediate corresponding tables for data in different IS.

Figure 2 shows the organizational structure of MUE "Vodokanal" of Irkutsk city. To date, more than 30 workspaces (marked with hatching) have information support from the ICE "ANGARA". These are all specialists related to the technological process of water supply and sewerage.

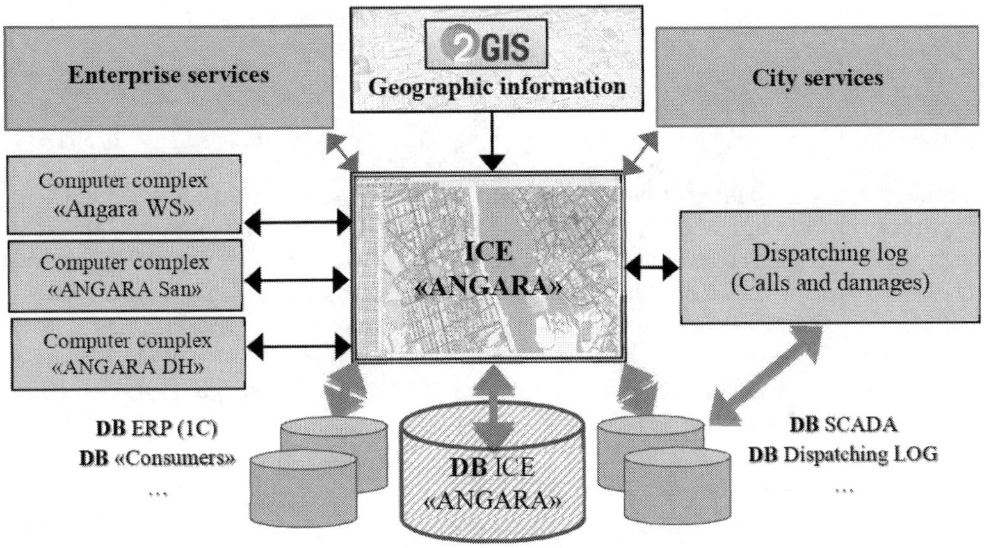

Fig.1. Organization of the CISE based on the ICE "ANGARA".

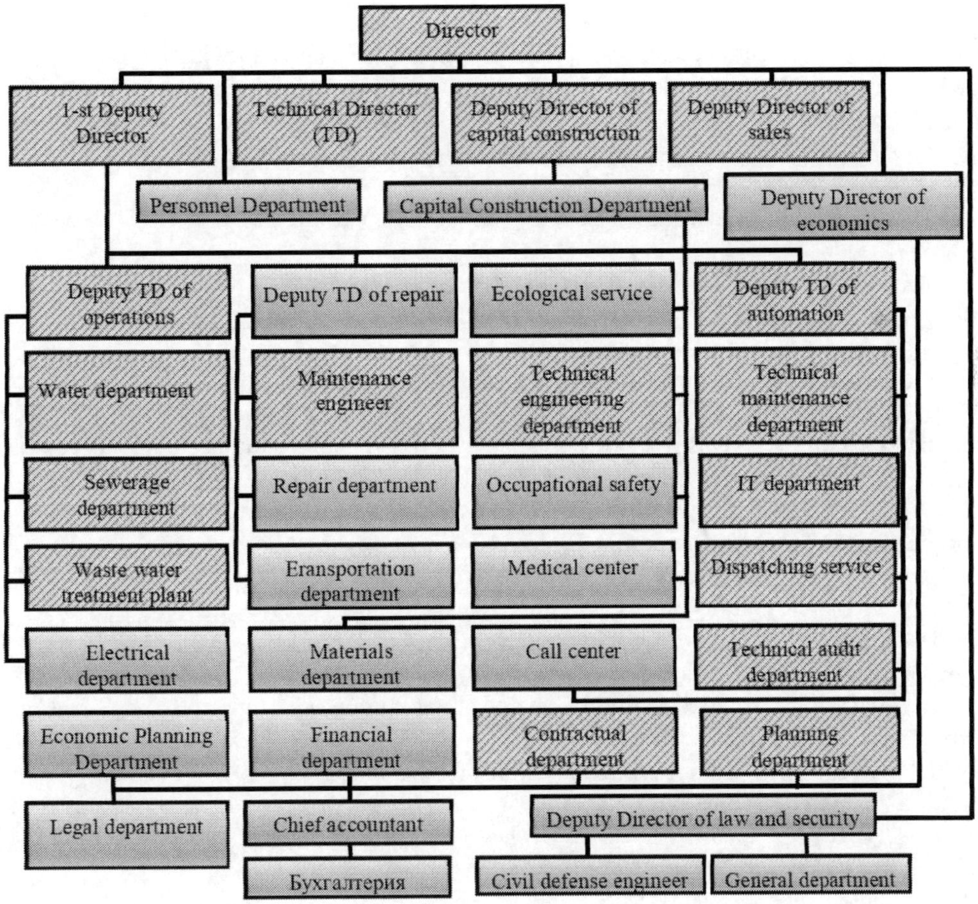

Fig.2. Organizational structure of the Municipal Unitary Enterprise "Vodokanal", Irkutsk.

Despite the fact that each department solves their own problems, all employees have access to a common, up-to-date information. For each employee, their access rights and the format of displayed information are asserted in accordance with its production tasks. In addition, a set of analytical functions of ICE "ANGARA" that simplifies the perception of large amounts of data is available to users.

Conclusions

The problem of "focal" automation of operating enterprises and the urgency of the CISE developing are disclosed. The definition, requirements, structure and methods of CISE organization are given. The most popular software tools for the CISE creating are presented. All these software products primarily automate business processes and give insufficient attention to informatization of technological processes taking place in the PLS and do not meet the requirements set by the CISE. A brief description of the ICE "ANGARA" and a scheme of the CISE creating on its basis are presented. An example of the CISE created in MUE "Vodokanal" of Irkutsk, is given. The organizational structure of the enterprise shows workplaces using CISE, including all necessary information about network and software packages for calculating and analyzing hydraulic operating conditions of water supply system.

References

1. http://lib.sale/pravo-grajdanskoe-uchebnik/informatsiya-gosudarstvo-edinoe.html
2. A.V. Alekseev, N.N. Novitsky, S.Yu. Obuzdin Water supply and sanitary technique **11**, 54 (2017) [in Russian]
3. http://gridder.ru/technologies/computers/edinoe-informacionnoe-prostranstvo-predprijatija-mif-ili-realnost/
4. A.V. Ostroukh, D.O. Gusenitsa, V.B. Golubkova, P.F. Yurchik IOSR-JCE **16/02**, 31 (2014)
5. S. Oman, R. Leskovar, B. Rosi, A. Baggia Technical Gazette **24**, 1889 (2017)
6. https://www.automationworld.com/article/technologies/erp/plm-mes-erp-closed-loop-product-lifecycle
7. E.A. Melnik, M.M. Hyamyalyainen, S.V. Smirnova, M.Yu Yudin Water supply and sanitary technique **3**, 20 (2012) [in Russian]
8. F.Yu. Kasatkin, S.V. Kostilyev Water supply and sanitary technique **6**, 45 (2010) [in Russian]
9. A.V. Kritsky, D.C. Kargapoltsev, D.N. Brusnitsin, e.t.c. Water supply and sanitary technique **11**, 45 (2015) [in Russian]
10. S.V. Hramenkov Water supply and sanitary technique **10**, 3 (2011) [in Russian]
11. http://www.oracle.com/us/products/applications/ebusiness/overview/index.html
12. http://solutions.1c.ru/catalog/vodokanal/features
13. https://www.galaktika.ru/erp/funkcionalnost.html
14. N.N. Novitskiy, A.V. Alekseev the Eleventh International Conference Computing and control for the Water Industry. **3**, 901 (2011)
15. N.N. Novitsky, M.G. Suharev, e.t.c. *Pipeline systems of Energy* (2015) [in Russian]
16. http://51.isem.irk.ru/angara
17. A.V. Alekseev, N.N. Novitsky, V.V. Tokarev, e.t.c. *Pipeline systems of Energy*, 221 (2007) [in Russian]
18. A.V. Alekseev, N.N. Novitsky, E.S. Melehov Proceedings of Irkutsk State Technical university **6**,12 (2014) [in Russian]
19. A.V. Alekseev, N.N. Novitsky, Science bulletin of NSTU **2**, 26 (2017) [in Russian]
20. A.V. Alekseev, N.N. Novitsky, S.Yu. Obuzdin Water supply and sanitary technique **11**, 64 (2017) [in Russian]

Internet system for hydraulic calculations of pipeline systems by the methods of the theory of hydraulic circuits*

Egor A. Mikhailovsky[1,†]

[1]Melentiev Energy Systems Institute of Siberian Branch of the Russian Academy of Sciences (ESI SB RAS), Department of Pipeline Systems, 130, Lermontov Str., Irkutsk, Russia, 664033

Abstract. The paper is focused on a characteristic of an innovative technology for remote application of methods of the theory of hydraulic circuits for modeling the operation of multi-loop water and gas pipeline systems, using the Internet browser. The aims, objectives and implementation principles of the technology resting on the concept of object-oriented modeling of pipeline systems and client-server architecture of the distributed software are presented. The results of its application are demonstrated.

1 Introduction

Considerable energy conservation and enhancement of pipeline systems efficiency are impossible without proper organization of their operating conditions. This in turn requires advanced methods of mathematical and computer modeling for calculation, analysis and quantitative substantiation of the conditions.

To date many organizations of Russia (MESI SB RAS, Politerm Ltd [1], JSC "Potok" [2] and others) and other countries (Intergraph Corporation [3], Sunrise Systems Ltd [4], Bentley Systems Incorporated [5], etc.) have developed and continue to develop and adopt a great many software packages. They are however intended to solve a specific class of modeling problems (analysis, synthesis, control) in a specific area (design, operation, dispatching) and are applied to certain types of pipeline systems (heat, water, gas, etc.).

Based on the scientific area successfully developed by Melentiev Energy System Institute SB RAS, namely the theory of hydraulic circuits [6,7], the Institute has originated a unique arsenal of effective methods for mathematical modeling, calculation and optimization, that can be applied to a pipeline system of any type and purpose. In this context, there is an obvious contradiction between the potential commonality of these methods and ineffective doubling of investigations on their software implementation and adaptation to the specificity of application.

* The research was carried out within the project III.17.4.3 of the Fundamental research program of SB RAS (AAAA-A17-117030310437-4) with finance support of RFBR and the Government of Irkutsk Region in the framework of research project № 17-48-380021

† Corresponding author: egor.isem@mail.ru

The object of the studies is an information-computation technology that can be used to solve the problems of flow distribution in pipelines of different types on the basis of general calculation methods. The purpose of the research is to develop principles, implement and test such a technology.

The following objectives are set to achieve the purpose: 1) implement the methods for calculation of steady-state isothermal flow distribution in the pipeline system, that are independent of the fluid (gas) flow patterns in the elements of the system; 2) systematize the flow patterns and develop the principles of implementation of the models that are responsible for the specifics of these patterns and interaction with the methods for calculation of flow distribution of the entire pipeline system; 3) develop a software architecture providing the required information, computation and analytical functions within a single user interface for the pipeline systems of different types; 4) implement and test such software with orientation to the application on the Internet.

The following interrelated ways are proposed to reach the above objectives: 1) transition to the concept of object-oriented modeling of pipeline systems in order to separate the software components that implement general calculation methods from the components responsible for specifics of the pipeline systems or a problem solved [8]; 2) develop the methods of the theory of hydraulic circuits with orientation to this concept [9]; 3) develop the software based on the client-server technology ensuring independence of development, upgrading and application of certain components of the application without reprogramming the rest [10].

This will enable: 1) a multiple use of once-implemented general methods in different software packages, for different purposes, applications and for different types of pipeline systems: 2) remote application of these methods in local or global computer networks; 3) minimization of the software support and development costs; 4) acceleration of adoption of the most advanced new and effective methods for conventional or new tasks in pipeline system modeling.

2 The problem and methods for calculation of flow distribution in pipeline systems

A traditional model of steady-state isothermal flow distribution in pipeline system includes two analogs of Kirchhoff laws and closing relations (flow patterns):

$$\mathbf{Ax} = \mathbf{Q}\,,\ \bar{\mathbf{A}}^T\bar{\mathbf{P}} = \mathbf{y}\,,\ \mathbf{y} = \mathrm{f}(\mathbf{x})\,, \qquad (1)$$

where $\bar{\mathbf{A}}$ – $(m\times n)$ incidence matrix in the calculation scheme with elements $a_{ji} = 1(-1)$, if node j is initial (final) for branch i and $a_{ji} = 0$, if branch i is not incident to node j; $\mathbf{A}$ – $[(m-1)\times n]$ incidence matrix formed from $\bar{\mathbf{A}}$ by deleting one of the rows; $\mathbf{x}$, $\mathbf{y}$ – n-dimensional vectors of flow rates and pressure differences in the branches of the calculated scheme; $\mathrm{f}(\mathbf{x})$ – n-dimensional vector function with elements $\mathrm{f}_i(\mathbf{x}_i)$, $i = \overline{1,n}$, that reflect the patterns of the pressure drop depending on flow rate (flow patterns) in the branches of the considered scheme; $\mathbf{Q}$ – $(m-1)$-dimensional vector of nodal flow rates with elements $Q_j > 0(< 0)$ for source (sink) at node j and $Q_j = 0$, if node j – a simple point of branch connection; $\bar{\mathbf{P}} = \{\mathbf{P}, P_m\}$; $\mathbf{P}$ – $(m-1)$-dimensional vector of nodal pressures.

The task is to determine vectors $\mathbf{x}$, $\mathbf{y}$, $\mathbf{P}$ at specified matrix $\bar{\mathbf{A}}$, vector $\mathbf{Q}$, known form of $f_i(x_i)$ for $i = \overline{1,n}$ and set pressure at one of the nodes (P_m).

There are many known methods and algorithms for solving this problem based on model (1). The basic methods however, as shown in monograph [6], are the Newton-

Raphson loop and nodal methods. Both methods are based on the Newton method but with a preliminary decrease in the order of linear systems of equations.

For example, the nodal method suggests search for a solution in the space of nodal pressures and is reduced to the organization of process $\mathbf{P}^{k+1} = \mathbf{P}^k + \Delta\mathbf{P}^k$, and in its each k-th iteration the correction $\Delta\mathbf{P}^k$ is obtained by solving the system $\mathbf{A}(\mathrm{f}'_x)^{-1}\mathbf{A}^T\Delta\mathbf{P}^k = -\mathbf{u}_1^k$, where $\mathbf{u}_1^k = \mathbf{A}\mathbf{x}^k - \mathbf{Q}$; $\mathbf{x}^k = \psi(\mathbf{y}^k)$; $\mathbf{y}^k = \overline{\mathbf{A}}^T\overline{\mathbf{P}}^k$; f'_x – diagonal matrix of partial derivatives $\partial f_i / \partial x_i$, $i = \overline{1,n}$, at point $\mathbf{x}^k$; ψ – vector function inverse to f with elements $\psi_i(y_i)$, $i = \overline{1,n}$.

Thus, the computational procedure of the nodal method does not depend on form $\mathbf{y} = \mathrm{f}(\mathbf{x})$, which should only meet the requirements of monotonous increase to provide the uniqueness of a solution to the flow distribution problem [8]. However, the specific features of this method implementation depend on specifics of calculating $\mathbf{x}^k = \psi(\mathbf{y}^k)$ and f'_x.

A typical form of the relationships is: $\mathrm{f}_i(x_i) = s_i \mid x_i \mid x_i - Y_i$, where s_i, Y_i – hydraulic resistance and effective head that are considered as specified coefficients, and $Y_i > 0$ for active branches (simulating pumps or pumping stations) and $Y_i = 0$ for passive branches (simulating pipeline sections). Here we have explicit expressions $\mathrm{f}'_{x,i} = \partial f_i / \partial x_i = 2s_i \mid x_i \mid$, $\psi_i(y_i) = \sqrt{\mid y_i + Y_i \mid / s_i} \cdot \mathrm{sign}(y_i + Y_i)$.

The relationship where s_i implicitly depends on x_i are also often applied. We will exemplify this by the known Darcy-Weisbach formula [11] for head loss in the pipeline which specifies $\mathbf{y} = \mathrm{f}(\mathbf{x})$ $h = \lambda(V)lV^2/(d2g)$, where d, l – diameter and length of the pipeline, (m); g – free fall acceleration (m/s^2); $V = \mathrm{V}(x)$ – fluid flow velocity, (m/s); x – mass flow rate, (kg/s); $y_i = \rho g h_i$; ρ – density of a transported medium (kg/m^3). Coefficient of hydraulic resistance λ depends on the Reynolds number $\mathrm{Re}(V) = Vd/\nu$, where ν – coefficient of kinematic fluid viscosity, m^2/s, which is assumed constant for isothermal flow. There are a lot of formulas to calculate λ. They reflect the purpose of the pipeline, its type, material of inner cover, service life, flow conditions (laminar, transitional, turbulent), etc. In Russia, the most widely used formula is Shevelev formula [12] $\lambda = A_1[A_0/d + C/(Vd)]^{\alpha}$, where A_0, A_1, C, α – coefficients depending on pipe material and flow conditions. Let us generalize all known cases of formulas for calculation of λ, then the i-th passive branch of the considered scheme will have the form $\mathrm{f}_i(x_i) = \mathrm{s}_i(x_i) \mid x_i \mid x_i$, where $\mathrm{s}_i(x_i) = \lambda_i(x_i)8l_i(\rho\pi^2 d_i^5)$ – resistance of branch, because $V_i = 4x_i/(\pi d_i^2 \rho)$.

Expression for the derivative in this case has the form $\mathrm{f}'_{x,i} = df_i / dx_i = (2s_i + s'_{x,i}x_i) \mid x_i \mid$, where $s'_{x,i} = ds_i / dx_i = 32 l_i \lambda'_{V,i} / (\rho^2\pi^3 d_i^7)$. Calculation $\mathrm{f}'_{x,i}$ for a pipeline of any type and purpose requires only specification of $\lambda'_{V,i} = d\lambda_i / dV_i$. In this case, search for x_i^k can be performed iteratively [8].

The relationships for modeling a gas compressor unit (GCU), for example, cause the major difficulty. The simplest relationship has the form [13]

$$\varphi(p_1, p_2, x) = \beta_0 p_1 |p_1| - p_2 |p_2| - \beta_2 x|x| = 0, \qquad (2)$$

where β_0, β_2 – coefficients obtained by quadratic approximation of a rated characteristic for the compression ratio, p_1, p_2 – GCU inlet and outlet pressure, x – performance referred to inlet conditions.

In principle, in this case we can apply the technique of double iteration cycles. According to this technique, in the internal iteration cycle, the traditional flow distribution problem is solved by the traditional methods (Newton-Raphson loop and nodal methods) at fixed values of parameters (s, Y) of classical closing relations, and in the external iteration cycle, the values of these parameters are specified depending on the obtained values of flow rates and pressures. For example, relationship (2) can be reduced to a conventional form $y = sx\,|\,x\,| - Y$, by denoting $y = p_1|p_1| - p_2|p_2|$, $s = \beta_2$, $Y = (\beta_0 - 1)p_1|p_1|$. Thus, the technique of double iteration cycles is rather universal, however it involves a many-fold increase in computational time for the flow distribution calculation compared to the traditional case. In [9,14], the authors propose modifications of the Newton-Raphson loop and nodal methods that exclude the need for this technique to take into account random closing relations including those implicit with respect to flow rate and dependent on pressure.

According to the technology of object-oriented modeling of pipeline systems, the method of flow distribution calculation, which is of general significance, should be implemented in an individual object "Network model" as an internal method for calling program. The properties of this object will contain information on topology of the considered scheme and values of boundary conditions (consumption and pressure of the medium at nodes). It is proposed to include the algorithms for modeling the hydraulic circuit elements in the object "Models of elements" whose properties will characterize the parameters of elements according to their types (for example, length, diameter, material of pipelines, pump characteristics, etc.). Internal methods of this object will calculate $f_i(x_i)_i$ and $\psi_i(y_i)$. Moreover, each object should support mechanisms of initial data input, for example by calling corresponding methods with arguments, and output of calculation results. Thus, the calculation process will be reduced to preliminary parameterization of properties and interaction of these objects. This approach makes it possible to apply ready implementations of calculation methods for modeling hydraulic conditions of different types of pipeline systems only at appropriate adaptation of the object "Models of elements". Programming of the calculated objects is proposed for example in the form of shared program libraries on the basis of component object model technology, which will allow their use independently in the programming languages of the software.

3 Purpose, main capabilities and case study of the software application

The software "Internet Hydraulic Calculation System" (IHCS) represents an example of the final implementation of the proposed approaches and is designed to model the conditions of water and gas pipeline systems on the Internet through remote application of advanced methods of the theory of hydraulic circuits. Accordingly, the software offers the possibility of calculations at any time, at any place and for any number of consumers provided they have the Internet connection and a standard web-browser. It can be useful for design and operation organizations, scientific and research institutions, engineers, post graduates, and students. The software has an intuitively understandable user interface, simple menu and its own mechanisms for constructing and editing the network scheme and data, which considerably reduces learning and application time.

The software is based on a client-server architecture. Client side – graphical user interface (Fig.1) is implemented based on Microsoft Silverlight [15], a special software component intended for the development of programs operating in the web-browser environment.

The IHCS software solves the flow distribution problems for pipeline systems of a random structure (with a random number and allocation of sources, pumping stations, reservoirs, consumption nodes and other elements) and configuration (multi-loop, tree, mixed) with fixed [6] and non-fixed [16] loads. The library of the calculation modules in the IHCS software contains implementations of several different methods of solving the flow distribution problem both for classical (Newton-Raphson loop and nodal methods) and new ("chord" nodal pressure method [17], fixed point iteration method [18-20], modified Newton-Raphson loop and nodal methods). In this case, the sparseness is used to solve the system of linear algebraic equations in these methods, which in most of the cases reduces the total time of calculation by two orders on average [10]. It is planned to further expand the library as the existing methods evolve and alternative methods appear.

The user can choose not only the method for calculation but also a calculated dependence. For example, the software can apply the formulas of Shevelev, Shifrinson [21], Prandtl–Nikuradze [22], Colebrook-White [11] or Hazen-William (which is used to a greater extent in other countries) to calculate the head losses in a water system pipeline section.

The IHCS software has been in pilot operation in open access (see 51.isem.irk.ru) on the Internet since 2013. An analysis of the data on the number of calculations of real networks and their fragments that were performed by users from different cities (Fig.2) makes it possible to conclude that these is a considerable and growing demand for this development despite the fact that there was not special work on its promotion.

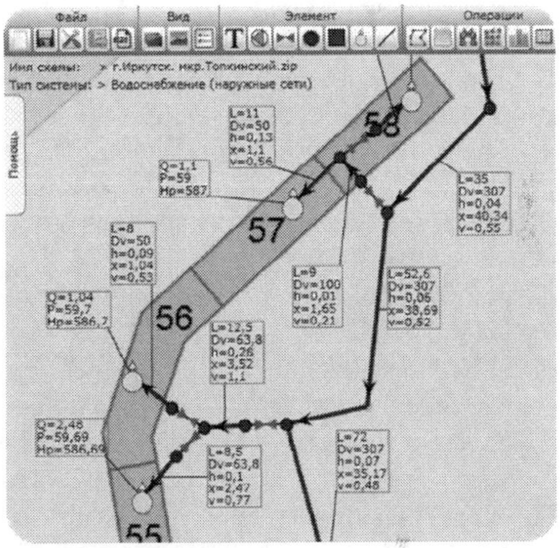

Fig. 1. Schema example.

Owing to the adopted computational capabilities [10], the IHCS software enables calculation for the schemes consisting of up to 1000 nodes which is correlated with the number of citizens of a populated settlement or a city district of about several tens of thousands of people. The time spent on the calculation is no more than 5 sec.

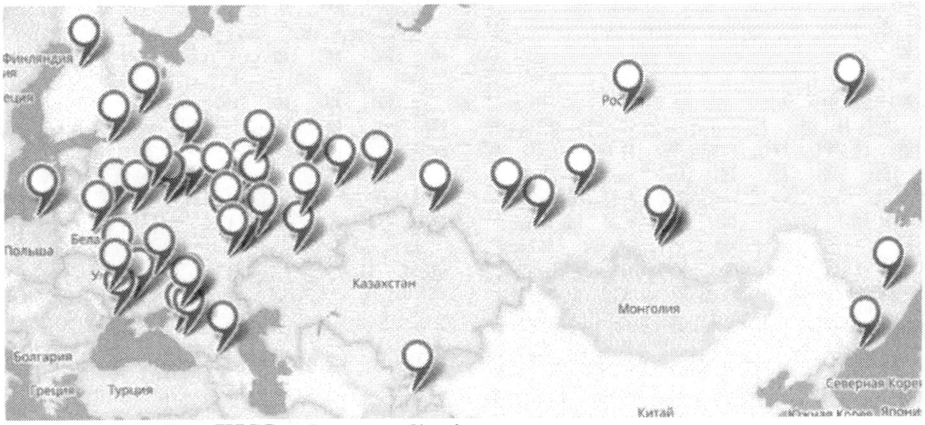

Fig. 2. Geography of the IHCS software application.

4 Conclusion

The paper demonstrates the IHCS software that represents an innovative development enabling for the first time a remote application of the effective methods of the theory of hydraulic circuits for modeling the hydraulic conditions of water and gas pipeline systems on the Internet. Until now, the IHCS software has been used to make tens of thousands of multivariate calculations for hundreds of hydraulic networks by a great number of the Internet users from tens of cities of Russia and FSU countries.

References

1. The site of the company Polytherm. [Online]. 2016. URL: http://www.politerm.com (accessed: 01/01/2016)
2. The site of the company Potok. [Online]. 2016. URL: http://www.potok.ru (accessed: 01/01/2016)
3. The site of the company Intergraph. [Online]. 2016. URL: http://www.intergraph.com (accessed: 01/01/2016)
4. The site information about the program Epanet [Online]. 2016. URL: http://www.sunrise-sys.com (accessed: 01/01/2016)
5. The site of the company Bentley [Online]. 2016. URL: http://www.bentley.com (accessed: 01/01/2016)
6. A.P. Merenkov, V.Ya. Khaselev. *Theory of hydraulic circuits* (Science, Moscow, 1985) [in Russian]
7. N.N. Novitsky and others. *Hydraulic circuits. Development of theory and applications* (Science, Novosibirsk, 2000) [in Russian]
8. N.N. Novitsky, E.A. Mikhailovsky. Bulletin of the IRSTU, **7**, 170-176, (2012) [in Russian]
9. E.A. Mikhailovsky, N.N. Nikolai St. Petersburg Polytechnic University J: Phys. and Math., **1(2)**, 120-128, (2015)
10. N.N. Novitsky, E.A. Mikhailovsky. Scientific bulletin of NSTU, **3**, 30-43, (2016).
11. A.D. Altshul. *Hydraulic resistance* (Nedra, Moscow, 1982) [in Russian]
12. F.A. Shevelev. *Tables for hydraulic calculation of water pipes: Ref. manual* (Bastet Ltd, Moscow 2007) [in Russian]
13. A.G. Nemudrov, V.I. Chernikin. Gas industry, **3**, 31-34, (1966) [in Russian]
14. N.N. Novitsky, E.A. Mikhailovsky. Energy Pipeline Systems: Mathematical and Comp. Tech. of Intellectualization, 51-59, (Science, Novosibirsk, 2017) [in Russian]
15. M. McDonald. *Silverlight 5 with examples in C # for professionals* (JSC "I.D. Williams", Moscow, 2013)
16. V.G. Sidler, S.V. Sumarokov, V.R. Chupin, Water supply and sanitary equipment. **2**, 4-5, (1989) [in Russian]
17. N.N. Novitsky. Proceedings of the Academy of Sciences, **6**, 56-69, (2013) [in Russian].
18. N.I. Baranchikova, S.P. Epifanov, V.I. Zorkaltsev, Water and ecology. Problems and solutions. **2**, 31-38, (2014) [in Russian]
19. R.T. Feizullin. Siberian Journal of Industrial Mathematics, **2(2)**, 176-184, (1999) [in Russian]
20. J. Krope, D. Dobersek, D. Goricanec. WSEAS / IASME International Conference on Fluid Mechanics, 59-62, (2006)
21. B.L. Shifrinson. *Basic calculation of heat networks. Theory and methods of the calculation* (Gosenergoizdat, Moscow, 1940) [in Russian]
22. V.I. Manyuk. *Installation and operation of water heating networks. 3rd ed.* (Stroiizdat, Moscow, 1988) [in Russian]

Improving of the quality and reliability of water supply systems on the basis of modern information and computing technologies*

Aleksandr V. Alekseev[1, †]

[1] Melentiev Energy Systems Institute of Siberian Branch of the Russian Academy of Sciences (ESI SB RAS), Pipeline Energy Systems Department, 130, Lermontov str., Irkutsk, Russia, 664033

Abstract. The article deals with the issues of increasing the reliability and quality of water supply systems operation on the basis of modern methods and software complexes for the analysis and development of hydraulic conditions. Against the backdrop of the analysis of the literature on the problem of reliability, the relevance and insufficiency of attention to maintaining the required level of reliability at the stage of water supply systems operation are revealed. The main factors that affect on operational reliability are considered. These factors are largely associated with the competent organization of operating conditions of water supply systems. A brief description of the «Angara-WS» computer program for solving the problems of analyzing and developing of hydraulic conditions, as well as the experience of its practical application, is given. A special feature of this complex is its universality, the possibility of multilevel representation of models, the execution of one- and multi-level calculations, integration into a common information space of the enterprise, automation of the processes of mode analysis, accumulation and analysis of damage statistics.

An analysis of literature and problem statement

In Russia there are more than 577 000 km of centralized water supply systems, about 60% of them have significant physical depreciation, one third of them requires urgent replacement. At the same time, only 1.6% of the total length of pipelines is replaced per year [1]. In these conditions, the maintenance of reliable and high-quality water suppling of consumers is complicated. A reliable water supply is understood as the ability of the water supply system (WSS) to provide consumers with water in the applied volume with the required parameters at any time. In the Russian normative documents there are no indicators of reliability of the WSS and their subsystems and also there is no a single interpretation of the WSS reliability term. The only requirement for the reliability of the WSS operation is the requirement of constant quality supply of all consumers with water.

* The research was carried out within the project III.17.4.3 of the Fundamental research program of SB RAS (AAAA-A17-117030310437-4)
† Corresponding author: asanya@mail.ru

However, construction rules [2] allow a short-term reduction in the supply of consumers in emergency conditions up to 30% of the volume of water supply.

The main causes of accidents in WSS that invoke a reduction of supply level to consumers are: the rupture of pipelines, damage of shut-off and connecting devices, disconnection of power supply to pumping stations and water intake facilities. There are following main types of pipeline damages: mechanical (construction works near water lines, design errors, soil subsidence), corrosive (for steel and cast iron pipes crack formation is typical), welded seam fractures (defective welding, pipeline deflection, temperature stresses). Each WSS has its own typical causes of accidents, depending on climatic and operating conditions, water quality, etc. On average, the frequency of failures in Russian WSS is in the range 0.5 to 2 accidents per 1 km of the pipeline per year [3].

The problem of WSS reliability arises during the design of systems and in their operation. Reliability of WSS operation is closely connected with the concepts of controllability and manufacturability.

Reliability of WSS during design can be improved by: network ringing, duplication of pipe sections, increasing their diameter, choosing the most suitable for these operating conditions of pipeline material, building backup water sources, installing backup and emergency pumping units, connecting pumping stations to various electrical substations, installation of reservoirs, emergency automatics, accounting for the planned prospective load of consumers, etc. This allows for the operation phase eliminate some of the accidents in WSS without limiting of consumers supplying.

The operational reliability of WSS is determined both by the reserves set at the design stage and by the operating conditions. As is known, the process of water consumption has a stochastic character, as a result of which the WSS constantly functions under changing parameters conditions (pressure, flow, concentration of disinfectants, etc.) and regimes. The reliability of WSS largely depends on the permissibility of operating conditions. Ensuring of permissibility of operating condition is observance of technological requirements of state parameters for each networks elements. The further the values of the mode parameters are from the boundary values, the higher the reliability of the system operation. Or, in other words, the permissibility of the operating conditions is a necessary condition for ensuring the reliability of the regime. Thus, the organization of hydraulic regimes has the greatest impact on operational reliability. In addition, quality and timeliness of repair and diagnostic works affect to the operational reliability.

The issues of increasing the reliability of the existing WSS are the subject of a large number of works, the main part of which deals with the selection of pipe materials, equipment and optimal terms of their operation. Il'in Yu. A. made a significant contribution to the study of reliability problems of WSS [4 and others]. Among modern publications, it is possible to single out the works of Galperin E.M. [5, etc.], which consider ways to improve the reliability of WSS through the use of polythene pipes as the most reliable and having a short repair time. In [6], the problem of increasing of the failures number in WSS in connection with the aging of equipment is considered and a technique for switching from an unsteady flow of failures to a stationary one is proposed, which makes it possible to use simpler engineering approaches to assess system reliability. The article [7] proposes a methodology for WSS reliability accounting in planning of the system development. In work [8] the technique of an estimation of probability of fault-free consumers supplying in with a glance to the probabilities of WSS failures and probabilistic character of water consumption is offered. This technique is based on the decision of a flow distribution with pressure-driven demand. The study [9] shows the dependence of the number of accidents on the pipeline, depending on its material, diameter and average pressure. In work [10] the technique of calculation of WSS reliability and quality indicators on the basis of hydraulic calculations on the aggregated model is offered. For a number of territories, seismic activity is one of the significant factors affecting to the WSS reliability. The account of seismic effects on the WSS reliability is given in [11, 12]. In all mentioned works, the reliability assessment of a particular system is reduced to multivariate hydraulic calculations and the need to develop a mathematical model or programs for hydraulic modeling (or adaptation of existing products such as EPANET [13]). At the same time, the authors do not pay enough attention to the permissibility of the regimes, confining themselves to checking of the level of consumers supplying. Thus, without using high-quality universal computer programs for modeling the hydraulic regimes of WSS operation, it is impossible to estimate their reliability exactly.

Brief description of the information-computer complex "ANGARA-WS"

Software for automating development of operational conditions for various types pipeline systems has been creating and implementing in ESI SB RAS for many years [14]. For the WSS simulation, the information-computer complex (ICC) "ANGARA-WS" (actual version 1.5) was developed, consisting of the information-computer environment "ANGARA" [15,16,17] and a set of calculation modules. The graphical interface of the ICC “Angara-WS” is shown in Fig.1. The main purpose of the ICC "ANGARA-WS 1.5"

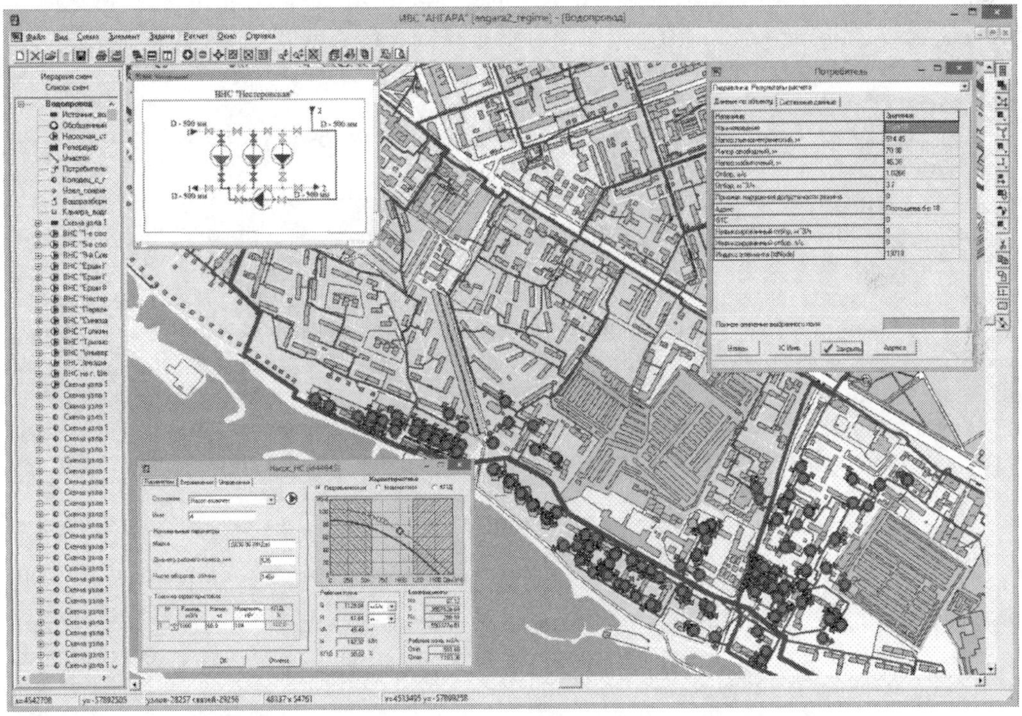

Fig. 1. Interface of the ICC "ANGARA-WS".

[18,19] is to automate the processes of operational conditions development (normal, emergency, post-emergency) of water supply networks of arbitrary configuration (branched, multi-loop), structures (with any number and location of sources, pumping stations, reservoirs, pressure and flow regulators and other elements) of any dimension. The program allows to organize the calculation scheme, both in single-level and in multi-level form [20]. At the same time, it is possible to carry out both the calculation of a single level (for example, main or distribution network) and a multilevel calculation of the WSS regime

as a whole. ICC makes it possible to perform: the calculation of the flow distribution with fixed demand (analysis of the network capacity at given loads of consumers); flow distribution with pressure-driven demand (analysis of the degree of consumer delivery at given network structure and characteristics); calculation of the operation conditions of pumping stations, piezometric graphs, permissibility conditions analysis.

Experience of ICC "ANGARA-WS" application in MUP "Vodokanal" Irkutsk city

Municipal unitary enterprise (MUE) "Vodokanal" operates water supply and sewerage networks in Irkutsk. The length of water supply networks on the balance of the MUE "Vodokanal" is 774 km; of sewerage network is 759 km, as well as 14 reservoirs, 22 water-pumping stations and 1794 fire hydrants. Irkutsk is located in a zona of high seismic activity and characterized by a complex terrain (geodetic marks of consumers are in the range of 420-510 m). The water supply network is divided into 17 zones with reduced or increased pressure [21].

The ICC "ANGARA-WS" is the basis of the automated working station of the operation conditions calculation engineers. The introduction of ICC in the group for conditions calculation in MUE "Vodokanal" in Irkutsk allowed:

1) to create and maintain a common database and an electronic calculation model for Irkutsk city. This model consist of water supply networks schemes (about 35000 nodes), network structures schemes (more than 10000 schemes) and sewage networks schemes (more than 65000 nodes);

2) to organize the access of various departments and services of the enterprise to this information;

3) to integrate the ICC into the common information space of the enterprise [22], that made it possible to avoid duplication and inconsistency of information in various departments of the enterprise.

For the dispatch service, the dispatching log "calls and damages" was developed. It allows to accumulate and analyze statistics on the damageability of the WSS, identify specific factors and conditions, and then take these factors into account in the design and development of WSS. Using this statistic, it becomes possible to calculate the probability of fault of system elements. Another important problem of the dispatcher's log is to maintain the database of WSS in the current state (by registering of all valves switching in the network).

After calibration of the obtained hydraulic model, a series of calculations were performed on it, analysis of which revealed:

1) the presence of a significant number of violations on the permissibility of the hydraulic regime, in particular, the excess of the maximum permissible head in the network; (consumers with excessive pressure, the required and actual characteristics of the pumping station are shown in Fig. 1);

2) the presence of closed circulation circuits around several pumping stations;

3) several closed valves in the network;

4) the most emergency parts of the network.

To increase the reliability and reduce pressure in WSS, the zoning scheme was revised, permissibility of hydraulic conditions was ensured. This allowed to stop 2 pump stations, to reduce pressure at others pumping stations and to determine the locations for pressure regulators setting. These events were held on WSS of Irkutsk in 2013-2014. These allowed significantly (about 25%) to reduce the fault rate in WSS in last years (Fig. 2a). In addition, it let to reduce the level of leakages in the network, the cost for electricity at pumping stations and improve the quality of WSS operation in general. The installation of pressure regulators allowed to reduce and stabilize the pressure in the most complex fragments of the network (Fig. 2b), that also positively affected on the WSS reliability.

Providing operational access to all employees of the enterprise to common information (including geo-information) allowed to reduce the time of registration and response on calls for damages and as a result the time of their repair. Analysis of statistics on the fault rate of pipelines showed that the greatest number of accidents fall on steel and cast-iron pipes. These accidents are caused by various factors: corrosion, subsidence, seismic action, etc. As a result, it was decided to gradually switch to polyethylene pipes. Recent years MUE has been working to replace the most emergency sections of the network with pipes from this material. For today, more than 144 km of pipes (19% of the total length) have been replaced [23].

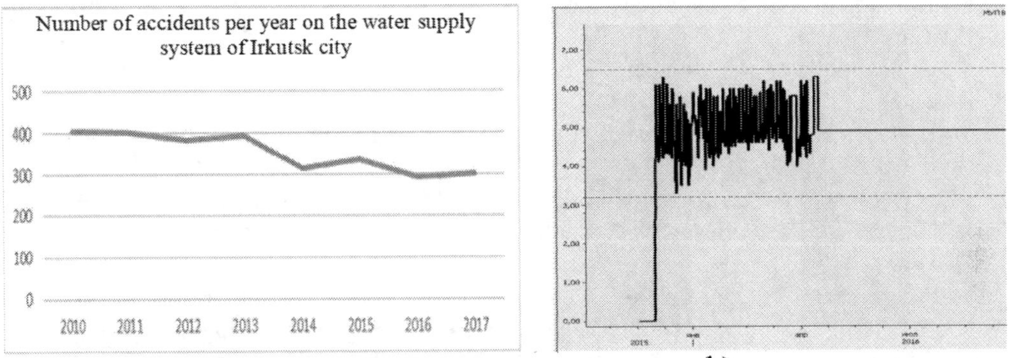

a) b)

Fig.2. Curve a) shows the number of accidents in Irkutsk WSS per year; b) displays the pressure change in one of network nodes before and after the pressure regulator installation.

ICC is also used for advance planning of repair work. Engineers calculate hydraulic conditions and work out all the variants of switching-off. The order of valves closure is made. The list of consumers with full and partial restriction of water supply is created. To facilitate the work, the ICC allows you to display only those node schemas in which the switching-off are made. After that, the network fragment is printed and sent to the management for agreement. Such a study of the repair plan on the electronic model also contributes to the rising of the system reliability.

Conclusions

The analysis of the problem of ensuring the WSS reliability shown that the problem of analysis and ensuring reliability requires time-consuming, multivariate calculations of the WSS hydraulic operating conditions in both design and operation.

The brief description of the new version of ICC "Angara-WS" is given. The feature of this ICC is the possibility of carrying out multi-level calculations of almost any dimension WSS, as well as effective analysis and maintenance of operational reliability of real systems.

The characteristic of the experimental and industrial operation of the ICC in the framework of the automated working station of the operation conditions calculation engineers is given. The main results obtained with his help is presented. The use of ICC for the analysis of the permissibility of the existing hydraulic regimes and the development of the events to ensure the reliable operation of the Irkutsk WSS is shown.

References

1. S.V. Khramenkov, O.G. Primin, About the reliability of water-bearing pipelines and the planning of their recovery (2013http://трубопровод.рф/статьи/все/2013/11/7/о-надежности-водонесущих-трубопроводов) [in Russian]
2. Building rules SP 31.12330.2012, *Water supply. External networks and facilities* (LLC ROSEKOSTROY, JSC Construction Research Center, Moscow, 2012) [in Russian]
3. M.A. Somov, M. G. Zhurba, *Water supply. Voluime 1. Water intake, supply and distribution systems* (ASV, Moscow) [in Russian]
4. Yu.A. Ilin, *Calculation of reliability for water supply* (Stroiizdat, Moscow, 1987) [in Russian]
5. E.M. Galperin, N.V. Shapkin, Water supply and sanitary technique, **8**, 28 (2013) [in Russian]
6. Yu.A. Yermolin, M.I. Alekseev, Water supply and sanitary technique, **5**, 68(2016) [in Russian]
7. O.G. Okeola, S.O. Balogu, Estimating a municipal water supply reliability (http://dx.doi.org/10.1080/23311916.2015.1012988)
8. V.R. Chupin, A.S. Dushin, Water supply and sanitary technique, **12**, 35 (2017)
9. V.S. Ignatik, S.V. Sarkisov, P.A. Putilin, The Eurasian Union of Scientists (EUS), **4 (13)**, 49 (2015)
10. M. Maglionico, R. Ugarelli, Reliability of a water supply system in the quantity and quality terms (http://studylib.net/doc/8739084/reliability-of-a-water-supply-system-in-quantity-and- qual ...)
11. M.B. Javanbarg, S. Takada, Seismic reliability assessment of water supply systems (http://www.academia.edu/5402910/Seismic_Reliability_Assessment_of_Water_Supply_Systems)
12. D. G. Yoo, D. Kang & J.H. Kim, Reliability Engineering and System Safety, **146**, 79 (DOI: 10.1016 / j.ress.2015.10.001, 2015)
13. https://www.epa.gov/water-research/epanet
14. N.N. Novitskiy, A.V. Alekseev, The Eleventh International Conference Computing and control for the Water Industry, **3**, 901 (2011)
15. http://51.isem.irk.ru/angara
16. A.V. Alekseev, N.N. Novitsky, Science bulletin of NSTU **2**, 26 **(**2017) [in Russian]
17. N.N. Novitsky, M.G. Suharev, e.t.c. Pipeline systems of Energy (Sciense, Novosibirsk, 2015) [in Russian]
18. A.V. Alekseev, N.N. Novitsky, S.Yu. Obuzdin, Water supply and sanitary technique, **11**, 64 (2017) [in Russian]
19. N.N. Novitsky, M.G. Suharev, e.t.c. Pipeline systems of Energy (Sciense, Novosibirsk, 2017) [in Russian]
20. N.N. Novitsky, A.V. Alekseev, O.A. Grebneva, A.V. Lutsenko A. V., V.V. Tokarev, Z.I. Shalaginova. Energy (https://doi.org/10.1016/j.energy.2018.02.070) (in press)
21. A.V. Alekseev, N.N. Novitsky, R.V. Chupin, V.R. Chupin, Water supply and sanitary technique, **11**, 26 (2017) [in Russian]
22. A.V. Alekseev, N.N. Novitsky, S.Yu. Obuzdin. Water supply and sanitary technique, **11**, 54 (2017) [in Russian]
23. A.V. Kurtin, Water supply and sanitary technique, **11**, 72 (2017) [in Russian]

E3S Web of Conferences **39**, 04004 (2018) https://doi.org/10.1051/e3sconf/20183904004
Mathematical Models and Methods of the Analysis and Optimal Synthesis of the Developing Pipeline and Hydraulic Systems

Integrated Systems for Commercial and Technological Control of Energy Resources Consumption in Urban Areas and Businesses in Russia

Roman Belousov[1] , *Eugene* Fiskin[1], and *Margarita* Fiskina[1]

[1]Irkutsk National Research Technical University, Russia

Abstract. The article contains a brief description of the operation principles information and measuring system (IMS) called "Kumir-Resource". It is compatible with different manufactures' equipment, so it can create a network using equipment that is already in place, which is versatile. It can also network with the existing SCADA systems to provide them with necessary information.

In today's world, in order to use energy resources more efficiently there is a need to optimize the use of monitoring processes and managing supply chain in real-time. In some instances, there is also a need to add systems of energy resources production to the cycle.

In general, the real-time management is not an issue anymore. There has been a number of successful businesses developing SCADA systems for some time. However, in practice SCADA systems often require using specific and unique equipment, such as detectors, meters, etc. Oftentimes businesses already have similar equipment in place, so its removal, purchasing and installation of the new equipment can be expensive and may take time. There are cases when one business would have several SCADA systems in place, each would have its own task, but all of them would require the same information. Supplying this required information is done by installing several sets of detectors in the same control spot, which results in excessive equipment.

It is obvious that despite effectiveness of the real-time management of modern SCADA systems, there are still issues of information flow, namely:

- SCADA may not be compatible with the existing equipment;
- Different SCADA systems may have issues with proper information flow between each other.

Information and measuring system (IMS) may solve all these issues. It is compatible with different manufactures' equipment, so it can create a network using equipment that is already in place, which is versatile. It can also network with the existing SCADA systems to provide them with necessary information.

For several years, the STC "Kumir" has been developing IMS called IS "Kumir-Resource" that implements the above-mentioned approach.

The following approaches are used in the IMS:

- Network of batch data transmission and format batches of the measuring devices without transformation is used. Thus, the terminal devices should be a gateway. An abstract query system is used allowing for creating a system regardless of the type of the metering device.
- Data entry/output and system management is done via web interface to create internet-oriented metering system.
- BSP system architecture is used to design the system on the whole, which makes the service affordable for businesses in Russia; this will facilitate introduction of internet-based solutions to the housing and utilities companies.
- Communication with the metering stations is done using GPRS technology.
- Access to multiple devices. This allows us to query a large number of energy metering stations at a time.
- Data collected form the metering stations is stored and analyzed on a server, which makes it possible to store and analyze a big volume of information. The operator can easily retrieve the required data.
- Own web server is available to report to all interested parties via Internet.
- Data backup. In case the server fails, all data is preserved, which is hard to implement if every user keeps their own information locally.
- IMS "Kumir-Resource" is built on client server technology; client applications are web browsers, and web server is the system's interface.
- Cross-platform implementation: IMS "Kumir-Resource" is based on a Software-as-a-Service (SaaS) model. SaaS model means that software is not purchased, but it is rather part of service. The benefits of SaaS include:
- no system administration expenses;
- no setup required
- no support and renewal required

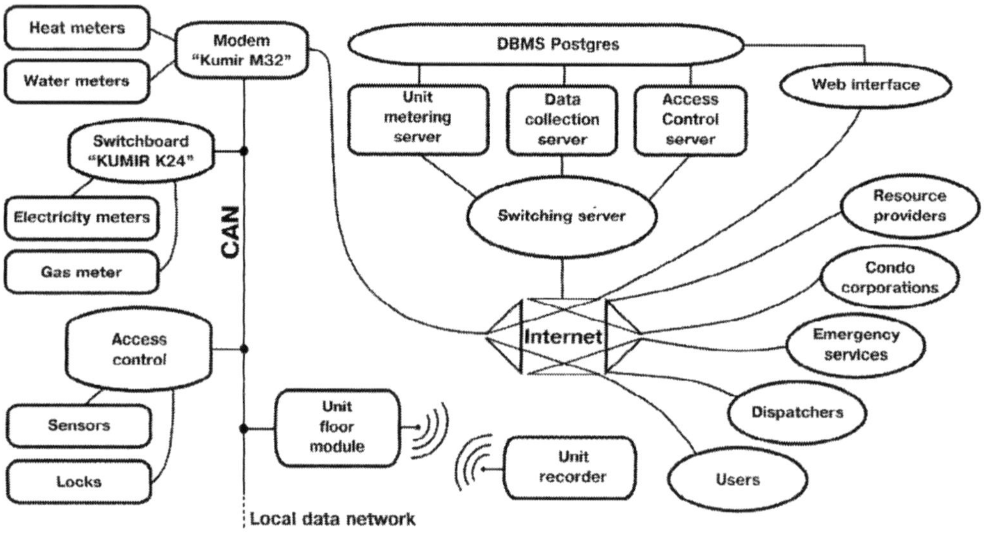

Fig. 1. Illustrates the architecture of the IMS "Kumir-Resource".

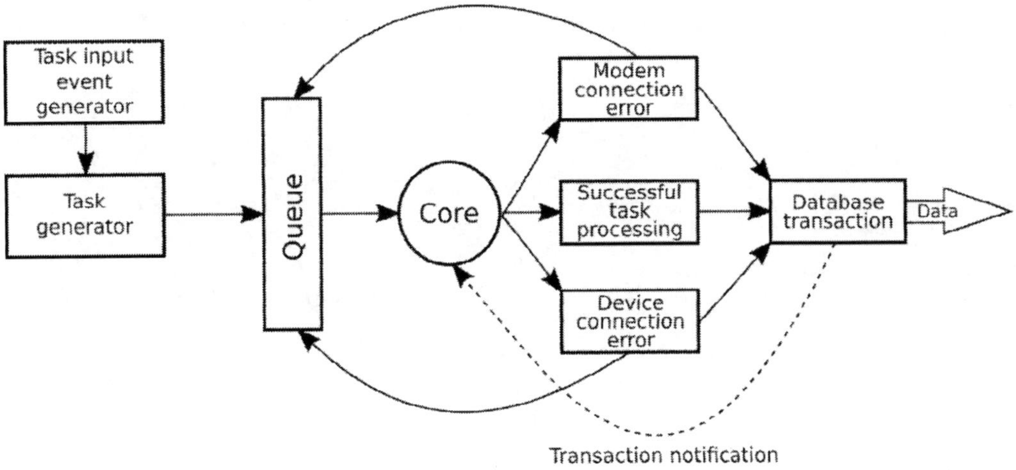

Fig. 2. Illustrates the functional scheme of the IMS.

In order to determine the efficiency of the system, an imitation model was built. One of the most effective and widely available complex discreet system modelling languages is the GPSS language. It can be successfully utilized in modelling of the systems formalized as systems of queuing. As language objects, the analogues of such standard components of queuing system used as requests, serving devices, queues, etc. Such a set of components facilitates creating complex imitation models while keeping the regular queuing system terminology. This language was selected to build the imitation model.

As seen in Image 2, the model is non-linear and contains feedback branching. The main element of the system is the core. It should be represented at GPSS as a multi-channel device (MCD). This is because the logic of query management of heat meter is the same for any system's source code.

The main task in information and measuring system modelling is to calculate the capacity of the core that manages the querying of the heat meters and data entry in the database.

The results of the modelling are as follows:

- A number of channels in MCD (core) determines the capacity of the system in general and is limited by bandwidth of the main communication channel from mobile carrier to system's server and is 500 channels per 1Mb of the band.
- The system is immune to communication interference because modelling showed that connection errors even at the rate of 25% do not decrease the capacity.

The following experiment was made on the dedicated server: simultaneous querying of 1,000 detectors in 100 buildings on 100 streets of 100 cities was modelled. The server load was 18% thus demonstrating high capacity of the system and possibilities of its scaling.

The abstract architecture of the querying system and the versatile protocol "Kumir-Net" facilitate stable work with detectors, meters, etc. regardless of their internal architecture, i.e, the IMS does not require devices of a specific manufacturer which allows for building IMS on the base of the existing devices.

Along with sharing information with all interested parties by means of web-interface (browser), the IMS can be integrated with SCADA systems to provide them with the data from the terminal metering stations, in the required format as shown in Image 3. This is achieved by utilizing switching server which does not not only organize work of the IMS itself, but also transforms the identification data from query spots (such as meters, detectors, etc.) into tag sets used by SCADA system. Therefore, the switching server acts like a gateway that provides data exchange between SCADA and IMS. IMS "Kumir-

Resource" can work with more than one SCADA system simultaneously, which solves the issue of excessive installation of metering devices.

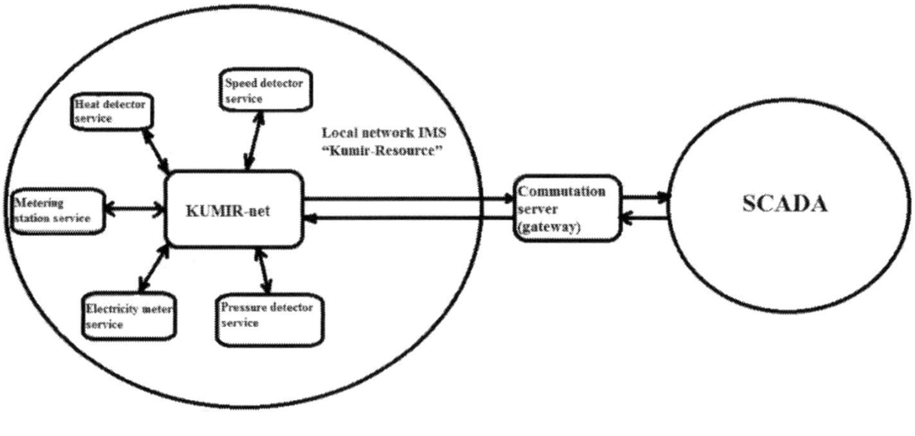

Fig. 3. IMS Integration with SCADA Systems:

References

1. N.N. Novitsky, M.G. Sukharev, A.D. Tevyashev and others, *Pipeline energy systems: Methodical and applied problems of mathematical modeling* (Science, Novosibirsk, 2015) [in Russian]
2. R.A. Belousov, I.A. Buzikov, N.N. Klimov, E.M. Fiskin and M.M. Fiskina, Modern technologies. System analysis. Modeling, **3**. 19 (2008) [in Russian]
3. R.A. Belousov, A.A. Pinkin, E.M. Fiskin and M.M. Fiskina, *EURO-ECO 2015. Das Internationale Symposium "Ökologische, Technologische und Rechtliche Aspekte der Lebensversorgung"*, 15 (Hannover, 2015)
4. R.A. Belousov, A.A. Pinkin, E.M. Fiskin and M.M. Fiskina, *EURO-ECO 2014. Das Internationale Symposium "Ökologische, Technologische und Rechtliche Aspekte der Lebensversorgung*", 28 (Hannover, 2014)
5. Jim Pinto. *Automation Unplugged: Pinto's Perspectives, Pointers, & Prognostications* (2004)
6. Terrence L. Blevins, Gregory K. McMillan, Willy K. Wojsznis, and Michael W. Brown. *Advanced Control Unleashed: Plant Performance Management for Optimum Benefit* (2003)
7. J. Rekstad, M. Meir and A. R. Kristoffersen, Energy and Buildings, **35** (2003)
8. B.K . Duncan, B.G. Bailey, Industry Applications, IEEE Transactions, **40** (2004)

AUTHOR INDEX